Negotiating Civil–Military Space

'In conflict zones across the world, militaries and civilian organisations are expected to work together but often struggle to do so effectively. Drawing on a range of historical examples, Marcia Byrom Hartwell's perceptive and empathetic analysis offers new ways of looking at this familiar problem. This is an important book that will be of use to scholars and practitioners alike.'

Gordon Peake, Australian National University

'Adept at navigating both the exclusive realms of international organizations and the U.S. Army, Dr Hartwell's ability to bridge the significant cultural gaps between military and civilian—and between academic and practitioner—is key to the book's insights. Many of the issues raised I witnessed first-hand in Iraq while serving with Dr Hartwell, as I was a primary facilitator of the transition of activities from the military to civilians. Addressing aspects of global security from military conflict to pandemics to crisis mapping to UAVs, her frank and balanced views and recommendations on civilian–military relations put this comprehensive book on the "must read" list for those interested in national security and humanitarian assistance, especially given the continued rise of non-state actors and the necessity of "whole of government" approaches.'

Jim Raimondo, The Fletcher School, Tufts University, USA

This book begins discussion at a point where many civil–military conversations end. Hartwell identifies underlying dynamics, key issues, and challenges that civilian and military organizations encounter when negotiating their roles in real and virtual volatile environments. These include managing expectations, understanding organizational missions and cultures, building trust, and exploring different approaches to violence. The impact of applied technologies on decision making processes and interventions is discussed in terms of recent and future complex crises. Linking earlier history to current discussions, this study makes an important contribution by reframing issues and outlining strategies to avoid unintended consequences and more effectively protect civilians in future operations. While geographic focus is on the Middle East, Africa, Central Asia, and Asia-Pacific, the core issues are applicable to negotiating civil–military relationships in a wide range of environments.

Marcia Byrom Hartwell's current research builds upon experience as an embedded civilian advisor with the U.S. Army in Iraq (2009–2011) and work as a 2013 CSCMO Scholar (Center for the Study of Civil-Military Operations at USMA, West Point) and 2011 Public Policy Scholar, Woodrow Wilson International Center for Scholars, DC. Dr Hartwell received her DPhil (PhD) from the University of Oxford.

Negotiating Civil–Military Space

Redefining roles in an unpredictable world

Marcia Byrom Hartwell

Routledge
Taylor & Francis Group

LONDON AND NEW YORK

First published 2016
by Routledge
2 Park Square, Milton Park, Abingdon, Oxon OX14 4RN

and by Routledge
605 Third Avenue, New York, NY 10017

First issued in paperback 2021

Routledge is an imprint of the Taylor & Francis Group, an informa business

© 2016 Marcia Byrom Hartwell

British Library Cataloguing in Publication Data
A catalog record for this book is available from the British Library

Library of Congress Cataloguing in Publication Data
Names: Hartwell, Marcia Byrom.
Title: Negotiating civil-military space : redefining roles in an unpredictable world / by Marcia Byrom Hartwell.
Description: Burlington, VT : Ashgate Publishing Company, [2016] | Includes bibliographical references and index.
Identifiers: LCCN 2015042277 | ISBN 9781472440457 (hardback) |
Subjects: LCSH: Integrated operations (Military science) | Civil-military relations. | Postwar reconstruction–Government policy. | Armed Forces–Civic action. | United States–Armed Forces–Civic action.
Classification: LCC U260 .H37 2016 | DDC 322/.50973–dc23
LC record available at http://lccn.loc.gov/2015042277

ISBN 13: 978-1-03-224249-1 (pbk)
ISBN 13: 978-1-4724-4045-7 (hbk)

DOI: 10.4324/9781315597645

Typeset in Times New Roman
by Out of House Publishing

Contents

Preface

I felt compelled to write this book from the viewpoint of someone who understands the questions that need to be asked rather than one who is able to supply all the answers. My goal has been to refocus civil–military conversations on issues that have prevented civilian organizations and military forces from identifying and developing more effective strategies to deal with multi-layered crises that are emerging from random combinations of violence and disasters. This book explores the early effects of social media platforms, communication technologies, and interaction of more recent technologies such as drones, robots, big data, and artificial intelligence on these issues, and why their technology enabled data are no substitute for human understanding and knowledge. The readers I primarily had in mind while writing this book belong to the international civil–military community. They are members of the US and international military forces and civilian humanitarian, aid, and development organizations who find themselves working alongside each other in dangerous areas. I also had in mind those who work offsite planning and overseeing policies that will be carried out by civilian organizations and military forces and the thousands of digital and other volunteers driven by a desire to help but who may have limited knowledge about how complex crises can evolve on the ground. Individuals with a professional or casual interest in international politics, international development and those struggling to understand US civil–military issues during the past decade may find these discussions useful.

The concept for this book grew out of an unanticipated encounter between myself and the US Army in Iraq from 2009–2011 who were probably as startled as I was to find myself embedded as a civilian advisor in their midst. Having researched and conducted fieldwork interviews on international conflicts and development in the UK for a MSc, then DPhil (PhD), I understood the intricacies of geopolitics and violence but knew little about US military culture. My previous interactions had been mostly with international forces in the UK whose members casually mingled with civilians in academic and everyday life. I was interested in the role of security forces during conflicts and as peacekeepers but they rarely allowed me to formally interview them especially in Northern Ireland, Serbia, and South Africa where I asked former combatants and community members for their opinions on forgiveness, revenge, and reconciliation. It was during this time I came to appreciate the ambiguity of supposedly clear cut motivations for conflicts and the vicious cycle between victims and perpetrators. I began to understand how easily people could be motivated to fight and how much more difficult it was to persuade them to stop. As an American living in the UK for nearly 10 years I had been the recipient of unfiltered news and comments that resulted in developing good debating skills and a somewhat Euro-American cross-cultural civil–military view. The basis for many discussions in the following chapters is drawn from this early research and experiences which strangely prepared me for working inside US military culture.

As an embedded member of US Army teams focused on reconciliation, then on military to civilian transition, several revelations about civil–military relationships emerged. One is how civilian humanitarian workers and military forces share parallel anxieties about their personal vulnerability in dangerous environments. Civilian workers worry about reactions triggered by military actions that could inadvertently turn them into targets while military forces worry that these same workers will not share information learned from local populations that could put them in danger. Another was realizing how differently civilian workers and members of military forces perceived violence. Unarmed civilians learn to closely monitor and evaluate potential threats and avoid them whenever possible while military forces are trained to move toward, engage, and neutralize violent events. This resulted in misunderstandings when civilians and military forces were making joint decisions on when to move outside safe areas. Finally, almost all efforts made by civilians and military forces to protect ordinary civilians from random violence in Iraq had limited effect. The best equipped army in the world and the most well intentioned civilian organizations were unable to stop bombs and attacks against people trying to go about their daily lives. One could see the cynical fatigue and resignation in their faces as the US rolled past their stopped vehicles in fast moving armored convoys. By all accounts a similar situation played out in Afghanistan. I suspect that the single greatest challenge to future negotiation of civil–military roles and operations will come from these targeted populations who are beginning to find their social media facilitated voice.

Acknowledgments

I want to thank the Woodrow Wilson Center for International Scholars, the United States Institute for Peace, and the Center for the Study of Civil–Military Operations (CSCMO) at the US Military Academy at West Point for providing an opportunity to research, think, and write in greater depth about these issues. My time as a Public Policy Scholar at the Wilson Center helped shape my views on protecting civilians that provided the basis for Chapter 4. As a CSCMO Scholar I traveled abroad to conduct interviews, give presentations for feedback, and update research that defined core issues on negotiating civil–military roles and relationships in writing Chapter 3. An opportunity to return to Iraq in 2012 while researching earlier US Institute of Peace programs provided me with updated perspectives and insights following the US withdrawal.

Special thanks also goes to Jim Graham, a retired US Agency for International Development (USAID) official for sharing important insights into civil–military dynamics during implementation of development projects and for suggesting I contact Stuart (Charles) Callison, a Senior USAID Development Economist who served both as a decorated US Air Force Liaison Officer to the Vietnam Air Force and as a DC-based Economic Advisor in the USAID Office of Vietnam Affairs, Asia Bureau. His analysis during and after the Vietnam War contributed to the subsequently shelved 1975 Viet Nam Terminal Report kept alive by contributors who scanned the original typed report. This report provided a key unfiltered civilian account of the evolution of civil–military efforts in Vietnam that formed the basis for Chapter 1.

I am grateful to all my friends and colleagues who while unnamed here are remembered for our many conversations, commiserations, and explorations of a wide range of relevant topics. Kudos to the individuals in the US military, especially the Army, who patiently explained their culture to me and to the many international humanitarian and development workers who initially introduced me to the complexity of the issues explored in this book.

Finally, this book is written with heartfelt gratitude to the memory of my parents Henry Green Byrom and Rita Morin Byrom who overcame their occasional misgivings to unconditionally support me in all my endeavors.

Marcia Byrom Hartwell

Introduction

Distracted by the latest group of "new" threats spurred on by a hyperconnected world the international civil–military community is once again being tempted to ignore recent lessons learned in Afghanistan and Iraq. Giving the false impression of being somehow unique, these "new" dangers are easily linked to old issues. Similar to many challenges discussed throughout this book they counterintuitively require a combination of faster responses shaped by deep analysis. This will only be achieved by more not less civil–military interaction, a willingness to develop relationships built on mutual trust, and a joint effort to clearly define organizational identities and operational roles.

Civilian organizations have acknowledged that earlier humanitarian guidance has been insufficient for interactions with foreign or international militaries in complex environments. They have called for greater clarity on "key aspects" of civil–military relationships, especially the "principle of last resort," humanitarian guidelines for requesting military assistance in relief operations, and information sharing protocols (Metcalfe *et al.* 2012: 29). Military forces who have become more experienced working alongside civilians and who regularly consult with national and international organizations have experienced equal degrees of frustration while negotiating operational space with humanitarian, aid, development and other organizations. This has often been the result of unrealistic expectations by military forces of civilian organizations who operate with relatively tiny budgets coupled by a lack of awareness about ways in which military forces can overwhelm local civilian operations and marginalize civilian organizational relationships with the host government.

Accepting there is no "right" answer for every situation the following chapters seek to expand discussion beyond narrow national and international interagency exchanges to acknowledge differences in organizational missions, cultures, budgets, and language. They reframe familiar issues and explore how to build civil–military trust based on common goals, organizational strengths, and vulnerabilities that deter violence and provide better protection for civilians. Growing input by local and global voices into decision-making processes and policies that affect them are reshaping old top-down civil–military operations. Many of the issues discussed here have often been omitted during formal working groups, task forces, and training, but are well understood by members of military forces and civilian humanitarian, aid, and development organizations operating in close proximity. Experience and knowledge are necessary to identify ways in which information and data flows analyzed independently from background or context are misleading. Understanding processes of violence and peace is crucial to identify tipping points between actions that can either provoke or diffuse violence in situations where good intentions could lead to unintended consequences, a civil–military dilemma depicted throughout this book.

It concludes by reviewing and offering insights into current civil–military challenges and ones that lie just ahead. New layers piled onto already complex crises—new armed and civilian actors, cascading disasters, greater emphasis on bottom-up policy development that crosses local/global divides, and new mixes of humans and technologies that make successful negotiation of civilian and military roles and relationships of even greater importance. The earlier stovepiping of humanitarian, aid, and security operations working alongside each other in the field has faded into the past. Rather than responding to each crisis as it occurs international civilian organizations and military forces will only be able to respond effectively by developing regional and global strategies based on relationships that respect differences, acknowledge limitations, and capitalize on strengths. They will need to develop more sophisticated civilian and military roles that manage risk, identify opportunities, and build organizations and structures that incorporate real time public input into disaster and crisis response. As new public voices influence ways in which civil–military roles and operational models are reshaped working groups and teams that combine a range of technological skills, knowledge, and multiple points of view will be necessary to effectively respond to future complex challenges.

Despite past difficulties, there is reason to be optimistic about negotiating future civil–military roles and relationships. Varying degrees of cooperation, coherence, and co-existence have been successfully demonstrated across the civil–military spectrum. A pragmatic attitude has emerged among international humanitarian and aid groups as former assumptions of operational neutrality have been undermined by terrorist and rebel groups who view them as pawns in a larger geopolitical agenda. Militaries realize that their role is limited. Perhaps most importantly as the value of macro/micro, fast/slow relationships and global/local relationships becomes better understood civil–military strategists and planners will find that revisiting lessons from their operational history has only grown more important in a world where there are fewer secrets and increased pressure for a quick response accompanied by greater potential for disastrous results.

This book is divided into three sections that describe the evolution of civil–military roles and relationships. Understanding the costs of recent and earlier civil–military lessons not learned is depicted in Part I: "Setting the stage". Descriptions of the reasons why the international civilian and military community can no longer afford this luxury of denial are outlined in Part II: "Reframing the issues". Part III: "Looking ahead" explores the "new" mixes of actors in wars, the impact of increased public input into developing policies, and the implications of human interaction with technology in this "new" civil–military space. Part I: "Setting the stage" begins by reviewing late twentieth century US civil–military experiences in Vietnam in Chapter 1 and how they set the stage for large scale civil–military operations in Afghanistan and Iraq. Chapter 2 explores US and international civil–military experiences in Iraq and Afghanistan and why these recent and earlier lessons from Vietnam are relevant to future interventions. Part II: "Reframing the issues" builds upon this history to question assumptions about civil–military roles and how to develop more effective responses to future challenges. The impact of newly emerging mobile communication technologies on these relationships and operations is an underlying theme in Chapters 3–5. Chapter 3 acknowledges fundamental differences in organizational cultures and missions and suggests how positive relationships can be constructed. Chapter 4 discusses why protecting civilians is difficult and Chapter 5 analyzes how military forces and civilian organizations differ in their understanding and reaction to processes of violence and why this contributes to unintended consequences. Part III: "Looking ahead" draws together earlier discussions to explore the implications of current and future multi-layered crises for the international civilian and military community. Overviews of actors in the "new" wars are discussed in Chapter 6. Chapter 7 discusses how the new local/global lightning quick connections facilitated by social media are reconstructing a bottom-up approach to policy development and Chapter 8 explores implications of the "new" civil–military space where humans will increasingly work with and alongside unmanned aerial vehicles (UAVs) or drones, artificial intelligence programs, and mechanical robots during emergencies, disasters, and complex crises.

Chapter 1: Winning "hearts and minds" in Vietnam revisits the original US "hearts and minds" efforts under the 1967 Civil Operations and Revolutionary Development Support (CORDS) program, which drew their inspiration from a similar British program in Malaya. Setting a precedent for civil–military operations in Afghanistan and Iraq, many of the civilian and military activities described were detailed in an unpublished 1975 Viet Nam Terminal Report by the Asia Bureau, Agency for International Development. This detailed multi-volume report describes the US and Vietnamese government "politicomilitary" security and development activities in Vietnam from 1954–1975 from a rarely heard civilian point of view. Written the same year that the US was forced out of South Vietnam it explores the trajectory and evolution of US civil–military roles as they changed from clearly defined separate assignments for civilian humanitarian and aid workers and military advisors to a combined massive "unity of effort" under the formation of CORDS during combat war conditions (USAID 1975a, 1975b, 1975c). The success of this effort was questioned at the time but the national trauma associated with the war resulted in both the US military and government organizations ignoring any lessons that may have been learned.

Chapter 2: "Unity of effort" in the long wars gives an overview of civil–military experiences in Afghanistan and Iraq by tracing increased military involvement in humanitarian and development activities in both countries. This chapter explores how civil–military roles evolved in Afghanistan and Iraq. As US civil–military missions drew upon stabilization and reconstruction policies from assumed successes they implemented Vietnam era "unity of effort" and "whole of government" crisis management frameworks that included comprehensive, coherent, and 3D approaches integrating pacification, reconstruction, and peace-support operations in each country. Unlike Iraq, Afghanistan had a longstanding limited international civilian humanitarian and development presence prior to the 2001 US invasion that was supported by a range of international civilian organizations and military forces. The 2003 US invasion of Iraq resulted in a primarily US implemented "hearts and minds" counterinsurgency strategy that assumed violence was driven by government neglect and poverty that could be stopped by civil–military programs, which improved governance and increased economic opportunities in each country (Jackson and Haysom 2013: 9). As both wars evolved US-led "unity of effort" and short term goals focused on winning the public's approval for their government while laying the groundwork for long term political, social, and economic changes. This concept was problematic for many international humanitarian and aid organizations concerned that this highly integrated civil–military approach would increase the vulnerability of their civilian staff and the populations they served (Jackson and Haysom 2013: 9–10). As large scale civil–military operations downsized in both countries it was widely acknowledged that civil–military dialogue needed to continue. Despite perceived failures the US and international civilian and military communities need to clearly assess recent experiences in Afghanistan and Iraq so that future civil–military efforts will be based on solid and more effective policies.

Chapter 3: Negotiating space begins Part II by describing characteristics that define and shape civil–military roles and relationships in complex crises and emergencies. These include fundamental differences in organizational cultures and capacities, timelines of missions and goals, budgets and material resources. This chapter emphasizes that building civil–military trust with the right leaders, relevant staff experience, and respectful personalities are key factors in successful implementation of all operations. Choosing visible and/or invisible interactions, understanding differences, focusing on common goals, developing creative problem-solving skills and inviting wider input into decision-making processes will be increasingly important. The early impact of "applied" information communication technologies and crisis mapping on civil–military relationships is explored. Many civilian humanitarian and aid organizations, peacekeeping, and military missions have not yet reconfigured their operations to cope with accelerated information flows that converge with events unfolding in real time. Understanding differences between raw data, factual information, and in-depth analysis will be critical as the lack of ambiguity or nuance in transmitted information makes data easy to manipulate and time consuming to vet and analyze.

This chapter emphasizes that building a deeper level of civil–military trust will be essential for the success of all future endeavors. This requires acknowledging the key role that leadership, personality, and experience play in successful outcomes, understanding threats from both civilian and military perspectives, and learning when and how to choose visible or invisible civil–military interactions. Past experience has shown that even when these relationships are tense the sooner civil–military dialogue takes place the greater the opportunity for civilian organizations to preserve humanitarian space and influence military conduct (Haysom 2013: 4). Accepting differences, focusing on common goals, developing creative problem-solving skills and inviting wider input into decision-making processes will facilitate effective, flexible responses in fast moving unpredictable technology-fueled environments.

Chapter 4: Protecting civilians discusses reasons why civilians remain unprotected in an era where information and videos depict human rights abuses in real time. Exploring the motivations of predatory governments who are often the greatest threat to their own citizens it reviews war ideologies that rationalize the governments' use of powerful militias, national security forces, and corrupt police to carry out assassinations, bombings, ethnic cleansing, forced displacement, genocide, and sexual violence against civilians (Wulf 2006: 29; Thakur 2010). Describing difficulties in identifying civilians who are innocent bystanders from those who pose danger to security forces and civilian workers it also acknowledges that civilians caught up in intense violence and its aftermath may be coerced or motivated by survival or ideology to aid enemies of security forces assigned to protect them. Opinions of local populations caught in the midst of territorial struggles between civilians and militaries in the day and informal militias, terrorists, and criminal gangs at night should be sought and incorporated into more effective protection strategies (Slim 2008: 34). In 2011 Afghans described that while development programs were welcomed, projects delivered by military forces or those working "under their auspices" potentially threatened a community's security as wherever international forces went, the Taliban would follow. International Security Assistance Force "high-profile," "quick impact" projects that focused on stabilizing violent districts failed to provide sustainable solutions for their humanitarian needs or to alleviate their poverty (Oxfam International 2011: 18).

Deepening civil–military communication will be increasingly valuable in developing effective civilian protection strategies in the future. Authors Beebe and Kaldor maintain that high tech weapons tend to have minimal effect in preventing the infiltration of criminals and militias in unstable environments and can sideline more important issues such as gaining the trust of local populations (Beebe and Kaldor 2010: 158, 7). Civilian protection will require more sophisticated analysis, evaluation, and operational techniques. Military forces could be trained to reinforce the fundamental link of being human by describing "non-combatants" as daughters, mothers, fathers, and grandfathers. Attributing qualities of humor, confusion, caring, fear, and resilience to local populations encourages military forces to exercise constraint, tolerance, and compassion in their dealings with civilians (Slim 2008: 34; Hartwell 2005). Making decisions to use "hard" power interventions or "soft" power public diplomacy to increase civilian protection requires greater awareness of potentially negative consequences especially when using social media platforms. Civilian and military interveners will need to combine nuanced observations, sophisticated analysis, with a mixture of high and low tech strategies that require focus, observation, understanding, coordination, and flexibility.

Chapter 5: Coping with violence discusses why understanding violence requires a different mindset and set of skills than learning about a country's culture and language. Building upon discussions in Chapters 3 and 4 violence is analyzed both traditionally in what Johan Galtung described as two critical issues—the use of violence and the ways in which its use is legitimized (Galtung 1990: 291) and within more recent frameworks that depict violence as a destabilizing health issue that prevents sustainable development and human wellbeing. Familiar but complex, violence can become easily enmeshed into national institutions and social fabric. Violent environments generally share two sets of characteristics, one with endemic violence lying just below the surface of daily life that appears to spontaneously erupt when the "right" combination of motives, circumstances, and timing converge, and another where urban and rural fighting between state and non-state forces takes the form of ongoing skirmishes and battles in the midst of civilians (Willman and Makisaka 2010: 46).

This chapter outlines several strategies with potential to meet these complex challenges and suggests that military forces have much to learn from unarmed international humanitarian and aid workers and local populations who have learned to survive in dangerous environments by correctly interpreting signals of impending violence and mixing a range of "hard" and "soft" security skills (Humanitarian Practice Network 2010: 112). Successfully deterring violence requires focus, observation, and an understanding of its characteristics. Actions that deter violence in one scenario may provoke it in the next. Successfully averting future threats will require the international civil–military community to better understand how state and non-state actors, public and private spheres, and internal and external actions interact to shape local violence. The role of mobile communication technologies in facilitating disinformation that instigates local violence and assists deterrence and survival by local populations is discussed in the context of balancing the need for short term mitigation with longer term prevention. Developing a commonly understood set of civil–military indicators and methods to identify the evolution of violence in future complex crises and emergencies will be essential to avoid unintended consequences.

Chapter 6: The new war challenge discusses the operational environment created by emergence of "new war" scenarios no longer fought within national boundaries but within regions of chronic insecurity where fast moving conflicts and internationally linked terrorism and crime have become the new normal. Mary Kaldor's earlier warnings about the "new" wars with their perpetually chaotic environments and no clear winners, losers, or negotiated settlements have become a global reality. No longer fought within national boundaries wars are evolving into regions of chronic insecurity where cyclical violence and recurring civil wars facilitate the formation of new militias, gangs, and financial networks that link transnational criminal activities with terrorist networks in mutually beneficial arrangements (Kaldor 2012; see Chapter 4).

This chapter describes the importance of monitoring areas of potential trouble and why identifying and understanding potential causes of complex crises is crucial in an era when random events and technologies can combine with natural and other disasters to instigate unpredictable disruptions and violence at any time and any place. Setting the stage for the final two chapters it defines trans-border regions of violence and their potential for accelerated destabilization through a random combination of natural disasters, environment and climate change, complex infectious health crises and pandemics, and the merging of natural disasters with technological developments to impact critical infrastructure and systems. Non-state actors and criminal networks take advantage of these situations to synchronize and/or propel regional conflicts onto the global stage for political or financial gain. It concludes by outlining challenges that these complex risks present to civilians, organizations and military forces and how new war scenarios will demand a reset of attitudes to develop a new more resilient civil–military mindset.

Chapter 7: New public voices explores how social media and related technologies are reversing the formation of political and security policies that were formerly developed and implemented from the top down. Local communities undergoing crises often become their community's "first responders" after an emergency, disaster, or security-related incident as they immediately launch rescue efforts and medical treatment that saves more lives than "second level responders" who arrive later (IFRC 2013: 73). These first responders can interact with the international civil–military community to identify and prioritize and communicate needs from the site. Beyond the emergency survival phase these local communities can also identify longer as well as short term needs that can be incorporated into external aid, development and security programs.

This chapter discusses the new interactive role and impact of local populations in developing more effective programs administered on their behalf. As the transnational "new" war environments continue to grow international humanitarian and development organizations have taken the lead in incorporating local perceptions into developing, implementing, monitoring, and evaluating projects and programs. It describes how the views of local populations including voices of marginalized and vulnerable populations are being sought by the United Nations in developing international policies and a post-2015 development agenda. The emergence of the digital humanitarians along with their use of social media and related technologies during disasters and conflicts and their impact on facilitating local voices who are increasingly voicing their opinions on the most effective short and long term security policies is explored. It concludes by suggesting how a flipped scenario where the effectiveness of conventional military programs and missions is increasingly challenged will shape future civil–military responses to rapidly evolving crises.

Chapter 8: Negotiating "new" civil–military space makes the case for looking ahead while emphasizing the value of drawing from historic lessons learned. This final chapter gives an overview of how humans and new technologies are mixing during complex crises. Civilian-driven UAVs better known as drones, which were once the sole province of the US military, are being increasingly utilized by a new generation of digital humanitarians to monitor human rights, environments, conflicts, disasters, and emergencies alongside a new generation of humanitarian robots who can assist in rescue, aid, development, and security scenarios during crises when human access is difficult or impossible. Reframing these conflicts, disasters, and emergencies into new combinations of complex crises it describes a "new" type of civil–military space where humans, cyber, aerial, and robot technologies respond and interact in unfamiliar ways.

This final chapter cautions how easily we can once again overlook the ways in which these "new" combinations of problems and challenges are often new versions of old problems. While humans operating in these environments will need to learn how to effectively manage and address the impact of communication, physical, and analytical technologies on shaping these human–tech relationships developing understanding between individual and organizational relationships will remain the bedrock for effectively responding to crises. Negotiating this "new" diverse discussion space will require utilizing a wider range of perspectives, ideas, and creative problem-solving skills than has been seen in the past. The importance of combining new thinking, skills, and old knowledge to develop adaptive, agile, nuanced responses from the members of the international civilian and military community cannot be over-emphasized. Tragic consequences from lessons unlearned or ignored continue to reverberate as the international community is challenged to identify ways in which current lessons are relevant to future crises. Despite the rapid growth of technology the importance of developing human relationships in negotiating roles and responses during complex chaotic crises will continue to underlie all future human and technological interactions.

Context and methodology

Material for this book has drawn from earlier research and fieldwork, personal experiences, and more recent interdisciplinary research, conversations, and interviews that explore the changing roles of armed forces in responding to humanitarian crises. Current discussions on humanitarian crises and ethics involved in addressing them, and local, regional, and international responses to civil–military actions have been included. The issues covered in Part II, Chapters 3–5 have been discussed in public presentations and conferences where they were publicly critiqued by diverse civilian and military audiences. Oral interviews documenting experiences of civilian and military practitioners have been treated with sensitivity and discretion. Names are used only when permission was specifically granted. The civil–military frameworks described in Chapters 2–6 build upon the author's highly interdisciplinary research, fieldwork, publications, and experience as an embedded civilian advisor on reconciliation and military to civilian transition for the US Army in Iraq (2009–11) and a subsequent visit to Iraq in July 2012 with the US Institute of Peace. Many core issues described in Chapters 3–5 were identified during a research trip to Northern Ireland to complete a MSc thesis paper on forgiveness in early post-conflict transitions in international development, London School of Economics, and during subsequent research and fieldwork on perceptions of justice, identity, and political processes of forgiveness and revenge in Northern Ireland, Serbia, and South Africa for completion of a D.Phil. (PhD) in international development, Oxford University.

This book differentiates between civilian humanitarian, aid, and development organizations that operate within a government's jurisdiction and ones that operate independently from governments. This is an important distinction as civilian government organizations operating under the same jurisdiction will have a different relationship with their national military forces than national and international civilian organizations who operate outside government control. However, these organizations are often less independent than assumed as they may be constricted by obligations to their financial donors, a widely discussed issue in international development that is touched upon here. While the author's military field experience has been primarily with US forces, civilians working with international forces confirm similar dynamics. Many topics discussed in this book are also relevant to domestic civil–military organizations operating within their own national borders.

Time and financial constraints have limited this comparative interdisciplinary research to a range of selected informal and formal conversations and semi-structured interviews conducted in person, by phone, Skype and email. Many conversations resulted in raising new issues, confirming others and in acquiring new or otherwise difficult to access research materials. Some interviews are anonymously cited when necessary. Qualitative analysis of information gathered from this process was primarily interpretive and insight driven, deriving a deeper meaning to identify processes and trends than that conventionally assumed from available information (see Alvesson and Skoldberg 2009). Empirical studies, observations, analysis, credibility of researchers and surveys, an understanding of the political and ideological contexts of the data gathered, awareness of the ambiguity of language, and the relationship between interviewer and interviewees were evaluated through an applied research lens. Initial literature surveys and interviews led to subsequent interviews, meetings, and refined searches of primary and secondary sources.

Part I
Setting the stage

1 Winning "hearts and minds" in Vietnam

The thread of many unresolved civil–military issues discussed here and in the remainder of this book can be traced back to US civil–military experiences and unlearned lessons during the Vietnam era. Lessons not learned from Vietnam led to misleading assumptions by the US military and government about the success of the "unity of effort" "hearts and minds" models that were the basis for expectations about international civil–military relationships in Afghanistan and for US civilian organizations and military forces in Iraq. This chapter gives an overview of US civil–military roles as they changed from separate assignments for civilian workers and military advisors after World War II to the merging of civil–military programs under the 1967 Civil Operations and Revolutionary Development Support (CORDS) attempt to win "hearts and minds" in Vietnam. It discusses how contentious this change was for US civilians implementing these programs and concludes with key lessons from Vietnam that continue to be relevant.

Developing the model

My introduction to a civilian viewpoint about civil–military roles in Vietnam was during a conversation with a US Agency for International Development (USAID) employee who had worked there and asked if I would like to see the unpublished 1975 Viet Nam Terminal Report (USAID 1975a, 1975b, 1975c). These typed manuscripts documented the USAID work in Vietnam from the beginning of US civilian and military assistance in 1954, throughout the Vietnam War, 1965–1969, to the fall of Saigon on April 30, 1975 (USAID 1975b: Vol. I, 4–5, 7/11). USAID employees had rushed to complete these reports by the end of 1975 only to see their efforts buried and ignored. Someone eventually scanned them to preserve as historic documents that described projects, personal observations and internal conflicts about the transition from civilian to military-led collaboration.

The civil–military model used in Vietnam was patterned after the successful 1945 Allied Marshall Plan named for US Secretary of State George C. Marshall that rebuilt infrastructure, the economy, and reformed European legal systems and governments following World War II (USAID 2014). Prior to US civilian and military assistance to Vietnam in 1954 a portion of the large scale military and economic assistance given to post-World War II France in the Marshall Plan was channeled to Vietnam, which was still a French territory (USAID 1975a: 3/96). This aid was administered by Allied forces who recruited civil administrators, city planners, urban development specialists, and administrators experienced with modern cities and governments to rebuild Europe (Cuny 1989). With few non-governmental organizations and the United Nations and international relief system in their infancy military forces were depended upon for support in humanitarian operations. As they demobilized Allied armies constructed refugee camps, distributed food, blankets, cots, conducted mass inoculations against typhus, typhoid, cholera, and donated military surplus to millions of displaced civilians. The 1947 Berlin Airlift solidified the military's humanitarian role when only airplanes were used to supply the entire city with food. The success of this operation contributed to the modern assumption that planes are necessary for delivering emergency relief, which in reality has proved to be an expensive often unsustainable mode of transporting supplies. A similar post-World War II era civil–military model was implemented in India, Palestine and other decolonizing countries as well as Korea, Greece, and Cold War flashpoint areas during the late 1940s (Cuny 1989).

During the 1950s Cold War era international assistance based on the Marshall Plan model expanded to include populations displaced by conflict and natural disasters inside newly emerging nations and as a tool in the power struggle between the US and Russia (Cuny 1989). The idea was that international civilian populations would choose democracy over communism if they were helped to develop democratic institutions and systems. President Harry S. Truman's 1950 Point Four Program formalized an international development assistance strategy with two goals, one was to create new markets for US business by reducing poverty and increasing production in developing countries, and the other was to reduce the communist threat by helping countries achieve prosperity through capitalism. From 1952–1961 programs that supported technical assistance and capital projects were the major form of US aid and important components of US foreign policy (USAID 1975b: Vol. I, 4–5). The earliest US aid to Vietnam was established in 1950 when the Special Technical and Economic Mission was created to provide assistance in cooperation with the French to their de facto colonies Associated States of Indochina including North and South "Vietnam." This early program helped to develop marine fisheries, civil aviation, health services, adult literacy, government information services and the Port of Saigon (USAID 1975b: Vol. I, 1/94). By 1954 Vietnam international aid had become an intrinsic part of American foreign policy as had the US view of itself as a benevolent force for preventing the spread of communism and supporting fledgling democratic countries.

Prior to USAID created by President John Kennedy in 1961, US international assistance was administered by three separate agencies, the Mutual

Security Agency, the Foreign Operations Administration, and the International Cooperation Administration (USAID 2014: History). Under USAID consolidation the 1960s became known as the "decade of development" (USAID 2014: History) when US economic aid to South Vietnam was unparalleled in scope and activities (USAID 1975b: Vol. I, 4–5). As operations increased relief agencies turned to UN international peacekeeping forces who had begun working alongside military forces to supply logistical and material support. This formed the basis for the US civil–military humanitarian, aid, and development model during the 1950s and 1960s in Vietnam.

US civil–military aid to Vietnam

Formal US civil–military aid to South Vietnam began after the 1954 Geneva Accords when the United States Operations Mission (USOM) was established to "plug the financial gap" left by the departure of the French and to assist South Vietnam's transition from a dependent territory to an independent state (USAID 1975b: Vol. I, 1/94). This followed a hundred years of conflict between the Vietnamese and their French colonizers, which had ended in a French defeat by pro-communist nationalists in the northwest corner of Vietnam at Dien Bine Phu (Brigham 2004). North and South Vietnam were temporarily divided at the seventeenth parallel with the understanding they would reunify after elections that would determine if they would incorporate under communist rule led by North Vietnam or by under a fledgling democracy in South Vietnam. Having recently ended the Korean War where they had successfully fought to protect the new South Korea democracy the US government and military were determined to stop the spread of communism in South Vietnam (USAID 1975b: Vol. I, 1/94).

In a series of multilateral agreements eventually known as the Southeast Asia Treaty Organization or SEATO the US attempted to push back against the threat of communist rule by focusing on nation-building (Brigham 2004) in a "South Vietnam" that now had increased their population by 8 percent with an influx of an estimated 900,000 refugees from the north between 1954 and 1955 (USAID 1975b: Vol. I, 3/96). From 1954–1958 US civilian aid projects focused on resettling these refugees and forming the new government. Included were efforts to reorganize and train a national army capable of maintaining internal security and resisting "external aggressors," rehabilitating and integrating over two million refugees into the economy, and assisting displaced populations and demobilized soldiers who might be vulnerable to "communist subversion." These aid projects were designed to make the government more responsive to the needs of the people and to increase popular support (USAID 1975c: 4–5). South Vietnam experienced relative security, financial stability, and fairly impressive economic development until 1959 when increased Viet Cong activity began to impact development programs. This activity was accompanied by budget deficits, inflationary pressures and a stagnating economy as instability grew to a point where the US civil–military presence took "a quantum leap forward" (USAID 1975c: 4/97).

Renegotiating civil–military roles

Civilian and military roles in Vietnam changed dramatically after violence accelerated in 1961. Prior to this time civilians administered humanitarian assistance and supervised economic and development projects, and military advisors embedded in the South Vietnamese military trained and built up South Vietnam's forces (USAID 1975b: Vol. I, 4–5, 12/43). As the US military launched large scale search and destroy operations in 1964 inhabitants of entire hamlets and villages were displaced. This resulted in large numbers of internal "refugees" (later called internally displaced populations) and pressure for jointly administered civil–military relief and rehabilitation programs. These included "temporary relief" that provided food, shelter, other items at official temporary government of Vietnam (GVN) camps throughout the country; "return to village," which assisted displaced populations who wished to return home to build new houses, community schools, dispensaries, roads, markets, wells, and buy food before new crops were harvested; "resettlement" to assist displaced populations establish new homes and communities in secure areas; and "in-place war victims" that helped those who had lost their homes and/or experienced death or injury of family members due to military actions and temporary displacement. Oversight of refugees was transferred from a provincial field program to a new USOM Office of Refugee Coordination in October 1965 led by an Assistant Director with refugee experience (USAID 1975c: Vol. II, 1/479, 3/481).

Responding to congressional pressure about the growing refugee crisis USAID formed the USAID Associate Directorate for Relief and Rehabilitation to liaise across civil–military agencies including USOM staff, the US military, GVN refugee offices, international organizations, and to provide technical advice to a new cadre of civilian provincial refugee advisors (USAID 1975c: Vol. II, 1/479, 3/481). The State Department then formed an interagency Vietnam Task Force that included the Defense Department, USAID, the US Information Agency and the Central Intelligence Agency (CIA) to reconsider US objectives. Two 1961 study missions inside Vietnam, one led by civilian Eugene Staley and the other by General Maxwell Taylor called for additional political and administrative reforms, new rural economic programs, increased US advisory and supply efforts, and developing and training a larger Vietnam military force. USAID was to implement rural economic programs and administrative reforms alongside military counterinsurgency programs operating inside hamlets, the lowest administrative level of provincial government.

USAID staff's reactions to USAID's role in these "short-term civil counterinsurgency programs" were sharply divided. Those opposed to this new civil–military collaboration designed to "stave off defeat" argued that their best contribution toward South Vietnam's stability would be to continue implementing programs designed to strengthen Vietnamese institutions and reinforce orderly economic and social development. They held the view that a direct response to the "guerilla crisis" was the responsibility of the Military Advisory Assistance Group (MAAG) and fell outside the traditional roles of the US Operations Mission and economic assistance programs (USAID 1975b: Vol. I, 12/43, 11/268). USAID staff in favor of working on counterinsurgency activities believed that unless the high level of violence was jointly addressed in the short term further development of long term institutions would be impossible. Internal conflicts intensified between 1962 and 1963 as debates on the appropriate level of USAID participation in counterinsurgency activities continued to be a primary internal issue. As US projects integrated with counterinsurgency strategies they evolved into partnering with military forces to develop a pacification program known as the Vietnamese Revolutionary Development. As the US and South Vietnam forces secured territory this civilian-led aid and development program would establish, expand and consolidate government control in villages by providing services, expanding economic programs, and lessening the impact of economic and social "consequences" resulting from military operations (USAID 1975b: Vol. I, 11/268, 5/20–7/22). Several long term economic stability programs such as USAID's Commercial Import Program continued but the major focus from 1962 onward was to achieve short term civil–military political and security objectives (USAID 1975b: Vol. I, 12/43–13/44).

The role of US forces had been evolving since 1960 when US Army civil affairs (CA) personnel were initially deployed on temporary duty assignments to help US Special Operations forces train South Vietnam's military (White 2009: 3). They were assigned to the III Marine Amphibious Force and 1st Infantry Battalion to assist thousands of civilians displaced by the fighting and to support the new Civilian Irregular Defense Groups (CIDGs) in Vietnam's Central Highlands. CA personnel built schools and taught modern agricultural techniques to improve the quality of life and encourage local tribes to fight the National Liberation Front (NLF) and the Viet Cong (VC) out of loyalty to the US military (Coffey 2006). In 1960 the US MAAG helped the Vietnamese Army organize and train 60 Vietnamese ranger companies in "antiguerrilla" tactics and prepared an organized pacification plan that served as a blueprint for later counterinsurgency/pacification efforts. As areas of instability broadened USAID and MAAG increased their staff. MAAG doubled its staff from 685 to 10,000 advisors and civilian-led USOM and MAAG had assigned advisors to each province by the end of 1962 (USAID 1975b: Vol. I, 11/42–12/43).

A CA unit 41st CA Company made up of 16, six man refugee teams was deployed in the winter of 1965. Each CA company included about 60 officers and 100 enlisted men with approximately 80 percent generalists and each CA unit was composed of one team leader (a captain/O–3 rank), a medical doctor, construction officer, a counter-intelligence officer and several CA "generalists." They often operated at 15–20 percent under full personnel capacity. CA missions had three key goals: eliminate the VC insurgency in South Vietnam, end VC recruitment in South Vietnam, and convince indigenous tribes living in South Vietnam to take up arms against the VC and the NLF in North Vietnam (White 2009: 3).

President Lyndon Johnson was determined to make the "pacification" of Vietnam into more than a military operation by winning the support of the local Vietnamese and building Vietnam into a model of economic, social, and political development in Asia. He envisioned this "other war" as a type of large US public works project that was a combination of the Tennessee Valley Authority and the Rural Electrification Administration on a national scale (Jones 2005: 105). President Lyndon Johnson declared "we must be ready to fight in Vietnam, but the ultimate victory will depend on the hearts and minds of the people who actually live out there" (Johnson quoted in Dixon 2009: 362). In 1965 the military advisory mission officially shifted to full combat status and integrated civil–military counterinsurgency programs were underway.

Winning hearts and minds

The concept of implementing a "hearts and minds" strategy to achieve national stability is often linked to Sir Gerald Templer, leader of the successful British counterinsurgency during the Malaya War (1948–60), who declared that military success was not achieved by "pouring more troops into the jungle, but in [winning] the hearts and minds of the people" (Templer quoted in Dixon 2009: 361). Templer interpreted winning "hearts" as gaining local emotional support and "minds" as responding to a population's "rational self-interest" (Dixon 2009: 363). He argued that influencing the political will of internal populations was critical to winning local support for national governments as communities would then be more willing to provide valuable information about insurgencies and less likely to join them. If the government in power inspired little confidence individuals would gravitate toward the insurgent group they deemed most likely to lead the next government. Templer believed these objectives could be achieved through the judicial use of "minimum force." Military forces would be expected to engage in social activities, civic projects, and psychological operations to reinforce a positive message and avoid alienating local populations. This would help reestablish a "cohesive" system of government rather than being solely focused on defeating an enemy (Dixon 2009: 358–9).

Templer's counterinsurgency approach caught the attention of the US. Prior to becoming commanding general of Military Assistance Command, Vietnam (MACV) General William C. Westmoreland visited Malaya accompanied by key US officials from Saigon and Sir Robert Thompson, head of the British advisory team to Vietnam, to better understand the chain of command that the British had utilized in their "hearts and minds" counterinsurgency strategy during that war. Westmoreland concluded that establishing a unity of command for advisors at the provincial level was essential for recreating a similar success in Vietnam (USAID 1975b: Vol. I, 23/280). Reinforcing this approach Robert Komer who would soon lead CORDS cautioned that providing more security in the short term would not guarantee victory in what was essentially a political war. Warning of potentially negative effects and the risk of increasing anti-Americanism in correlation with an increasingly visible military presence (Jones 2005: 108), Komer insisted that US pacification should primarily advise and support, stating "we do not intend to take over what the GVN and the ARVN [Army of the Republic of Vietnam] must do as an essentially Vietnamese task" (Komer 1970: 58).

American interpretation and use of the words "pacification and counter-insurgency" described a wide range of operations carried out by the US and Vietnamese (USAID 1975b: Vol. I, 2/259). "Pacification, counterinsurgency, rural construction, revolutionary development and civic action" were the most frequently used terms. In early years "counterinsurgency" indicating military operations against enemy insurgents and "pacification," which was an English language interpretation of an ancient Vietnamese term to describe actions that made a government more attractive to its constituents, were often used interchangeably. In later years English speakers used pacification to indicate military actions that sought to "gain control of a population" and the term counterinsurgency fell into disuse. "Development" referring to the process of improving government services in rural areas was frequently used by civilians in the earliest years of program assistance. The terms "revolutionary development," "rural development," "rural construction" and "civic action," were used at different points in time to refer to both military and civilian programs that could build stronger bonds between rural South Vietnamese and their government. USAID civilian advisors preferred the term "rural development" to describe USAID programs "whatever the stated purpose of these programs or whatever alternate terms were used to describe the various activities" (USAID 1975b: Vol. I, 3/260).

While US civilian and military province advisors worked together to implement these programs they continued to operate within separate chains of command and coordination of priorities had not kept pace with the intensity of the war (USAID 1975b: Vol. I, 23/280). American forces were increased to over 400,000 by 1967 as coordination and management problems intensified among civilian USAID field operations staff and MACV provincial and district advisors. Many Vietnamese provincial chiefs complained about Americans running around trying "to sell some program" while being pressured by their own government for similar reasons (USAID 1975b: Vol. I, 22/279). As growing numbers of US civilian and military personnel arrived to carry out pacification/counterinsurgency programs it became obvious that more effective coordination and oversight was urgently needed (Jones 2005: 105).

Unity of effort and CORDS

The concept of a "Vietnam czar" who would oversee all civil–military operations from his office in Washington DC was agreed upon in a January 1966 meeting between senior officials from civilian agencies, the Department of Defense, and the US Mission in Saigon but they could not decide whether this person would be assigned to the State Department, National Security Council or to the President's staff at the White House (Jones 2005: 105). President Johnson asked Robert W. Komer, a trusted National Security Council member and Harvard Business School graduate to explore this further. In his April 1966 evaluation "Giving a New Thrust to Pacification: Analysis, Concept, and Management" Komer reframed pacification into three separate initiatives that depended upon each other for success. The first would provide security for civilians, the second focused on disrupting the communist infrastructure to lessen its appeal while simultaneously implementing programs to win back popular support, and a third would implement pacification on a large scale. Komer felt that a massive effort would be required to turn the war around in favor of the US and South Vietnam. He acknowledged the lack of support for civilian-led pacification by both the US and Vietnamese military as they considered it essentially a "civilian problem" to be handled by GVN civil ministries, the US Embassy, Agency for International Development, and CIA, while they focused on the "main force war." The problem for civilian organizations was that most of the in-country resources—money, personnel, supplies, transport vehicles, ports, etc. were controlled by the military, and without military forces working with civilians to provide security all pacification efforts would be doomed to failure (Komer 1970: 52). Komer pointed out that the military's dominance in Saigon, the generally "weak and apathetic" South Vietnamese government, and US civilian agencies' lack of ability to operate at a pace required by the war hindered overall progress (Jones 2005: 108). Pacification and "other war" programs could contribute far more if US civilian-led efforts were prioritized over adding more troops (Andrade and Willbanks 2006: 13; Komer 1970: 57–8).

Ambassador Henry Cabot Lodge had begun restructuring civil–military efforts in Vietnam by early 1966. He appointed Deputy Ambassador William J. Porter to coordinate US support of rural activities and roles and duties within the US Mission. Porter formed a joint US Agency Planning Group to synchronize rural development planning between US agencies and the GVN's Ministry of Rural Development (MORD) (USAID 1975b: Vol. I, 24/281). They were consolidating project guidelines and prioritized areas for operations when pressured by Washington, Ambassador Lodge merged all USAID, United States Information Service, Office of Special Activities civilian field advisors and operations under a new Office of Civil Operations (OCO) led by one person in November 1966. Senior province officials were given the titles of USAID province representative and USAID's Wade Lathram became the first OCO director reporting to Ambassador Porter. This streamlined communications so that each USAID province chief needed to coordinate with only two others, the MACV provincial US military supervisor and the OCO supervisor about civilian activities in each province (USAID 1975b: Vol. I, 24/281).

President Johnson asked Komer to accelerate OCO's "unity of effort" giving him 120 days to accomplish this or be moved under the US MACV. Limited coordination with the US military and too little time to hire civilian staff left Komer unable to meet the deadline. Johnson subsequently dissolved the OCO in March 1967 and on 9 May 1967 signed a National Security Action Memorandum 362 assigning MACV with responsibility for pacification. He also created the position of Deputy in charge of CORDS with the rank of Ambassador reporting to MACV Commander General Westmoreland and promoted Komer to this new position (Andrade and Willbanks 2006: 9).

As Deputy for CORDS Ambassador Robert Komer now presided over the entire CORDS organization answering only to the MACV Commander. Four CORDS Deputies were assigned to the four military regions: 1st Military Region, Da Nang; 2nd Military Region, Nha Trang; 3rd Military Region, Bien Hoa; and 4th Military Region, Can Tho. The four Province Senior Advisors assigned to each military region answered to the CORDS Deputy in the military regions and the four District Senior Advisors answered to the Province Senior Advisors. Channels of communication were simplified with direction of regional and province advisors usually coming directly from the Deputy for CORDS, to the regional deputies, rather than from the Commander, MACV, through the commanding generals of the regions to the regional deputies and province senior advisors (USAID 1975b: Vol. I, 317).

All US civil–military activities on the regional, provincial, and district level including planning, programming, operations, evaluations, logistics and communications were now reorganized under CORDS' to speak with "one voice" to the Vietnamese (Komer 1970: 54–5). Under CORDS a civilian leader who managed military staff was in turn managed by a military officer. At the provincial level, military Province Senior Advisors had civilian deputies and civilian Province Senior Advisors had military deputies (USAID 1975b: Vol. I, 25/282–26/283). A typical province team would consist of a senior USAID representative assigned to be either the Deputy Province Senior Advisor or the Province Senior Advisor. A senior military officer, a lieutenant colonel or colonel was often assigned to one of those positions. One or more USAID public safety advisors, an agricultural advisor, one or more generalists, and occasionally a social welfare or refugee advisor employed by USAID completed the team. The US Information Agency supplied a staff information specialist and the Army usually assigned an engineer to act as an advisor (USAID 1975b: Vol. I, 25/282–26/283).

The primary goal for 1967 province teams was to implement a counter-insurgency program for rural Vietnam. They could draw upon additional technical assistance from the regional level where large, similarly constructed groups were established to link Saigon and field-based staff. This "sandwich" chain of command was in place at the district, province and regional levels up to the top of the US Mission, where the Deputy for CORDS, a civilian with ambassadorial rank, reported to the military commander of MACV, who in turn reported to the Ambassador. These widespread organizational changes significantly impacted Saigon-based American agencies and the Vietnamese government (USAID 1975b: Vol. I, 25/282–26/283). Having negotiated direct access to the President and fully supported by the Secretary of Defense Komer could effectively bypass the immediate chain of command when he chose, instantly making him one of the most powerful men in both the US and Vietnam (Andrade and Willbanks 2006: 9).

While Ambassador Lodge's successor Ambassador Ellsworth Bunker publicly declared CORDS to be a "unified civil military US advisory effort in the vital field of Revolutionary Development," which was "neither civil nor military but a unique merging of both to meet a unique wartime need" (Bunker quoted in USAID 1975b: Vol. I, 25/282) not everyone was pleased with this turn of events. Komer had been given the nickname "Blowtorch" by Ambassador Lodge who compared his earlier demands for faster progress to having a "blowtorch aimed at the seat of one's pants" (Jones 2005: 107). Komer's assessment of civilian efforts, which he called a "mess" in his pre CORDS fact finding Vietnam trip reports to the President, had infuriated several US civilian agencies. George A. Carver, CIA Special Assistant for Vietnam Affairs sent a memo to the Director of Central Intelligence saying that Komer was raising unrealistic expectations by giving the President the impression that under a unified effort pacification could be quickly accomplished with quantifiably measured results. He agreed with the need to establish village level security and eliminate Viet Cong influence but found a number of Komer's recommendations to be counterproductive, particularly his premise that additional resources and better management were the keys to winning the war. Carver's concerns reflected a deep and growing concern among civilian aid professionals about the militarization of pacification (Jones 2005: 109).

Between 1967 and 1972 CORDS attempted to simultaneously reconstruct Vietnam during combat operations. All USAID pacification activities and associated provincial, regional, and national level field personnel were incorporated into CORDS and were under MACV supervision by May 1967 (USAID 1975b: Vol. I, 5/20–7/22; 24/281–25/282). All US CA units that had been building CIDGs and other local militias were reassigned to CORDS by the end of 1967 (White 2009: 3). Remedying the Vietnamese government's lack of an agency capable of developing and managing rural programs US and Vietnamese counterparts bypassed traditional constraints to form the Rural Reconstruction Council to initiate pacification activities. The council was elevated to a ministry and the name changed to the Ministry of Rural Development (MORD). For unknown reasons Americans translated this as the Ministry of Revolutionary Development and the Vietnamese accommodated them by using that title in English (USAID 1975b: Vol. I, 15/272).

MORD was called a "super-ministry" by both supporters and enemies and given a streamlined fiscal system that allowed it to rapidly disperse funds with a minimum of constraints along with authority to provide hamlets with nearly any government service their inhabitants could desire.

With the backing of the Vietnamese Premier Ky and the US government, MORD set a goal to pacify two million inhabitants in 2,000 hamlets adding to the 7,842,000 in 3,620 of the country's 10–12,000 hamlets considered to be pacified under recent programs. Success of resettlement projects depended on establishing sufficient military security. Before 1968 the majority of MORD's employees were 40,000 rural development cadre team members who were expected to organize each hamlet, supervise construction projects, oversee elections of hamlet chiefs then move on to the next assigned hamlet and repeat the same. These militia-like cadres organized into 59 man teams who were armed and wore black clothing, and were constantly targeted by the VC who demoralized team members and destroyed their newly built structures. VC also warned local populations about unpleasant consequences if they cooperated in the future, resulting in slow progress for these efforts (USAID 1975b: Vol. I, 18/275).

The MORD approach changed dramatically after North Vietnamese conventional military forces simultaneously attacked Saigon and 30 of the 44 provincial capitals in what became known as the Tet Offensive on 31 January 1968. US civilian and military field advisors found themselves under mortar, rocket and ground attack and a number of USAID workers were killed and wounded. Repelled by US military forces, nevertheless this attack marked a pivotal moment for all US civil–military roles and pacification programs, and began what turned out to be a final phase of "Vietnamization" or Vietnam-led programs, which shifted US goals once again (USAID 1975b: Vol. I, 28/285). The new president Nguyen Van Thieu (as of October 1967) responded to the Tet Offensive by forming a Central Recovery Committee to coordinate and expedite recovery efforts for refugees, to restore cities, reopen communication between the national government and provinces, and develop a "national spirit of unity." The committee included all government ministries presided over by the President. In reality, working meetings were attended by representatives of each ministry and overseen by a Vietnamese lieutenant general. This group became the model for the new Central Pacification and Development Council (CPDC), which operated directly under the Office of the Prime Minister. With US backing this new program replaced MORD as the manager of rural activities and local military affairs, in effect becoming the Vietnamese counterpart of CORDS (USAID 1975b: Vol. I, 29/286–30/287).

The joint emergence of CPDC and CORDS resulted in the mutually developed Accelerated Pacification Campaign in November 1968. Vietnamese forces had demonstrated their ability to mostly hold their ground against immediate and heavy attacks by the VC when supported by US military forces. The goal was to secure more hamlets by simplifying and reducing the number of military, government, and political requirements. If judged successful it would be followed up with a long term development program accompanied by Vietnamese forces assigned to stations in less secure hamlets with orders to hold them at all costs. Favorably impressed by these results Vietnam and the US refined this concept into a 1969 pacification program that refocused pacification on the village level rather than on hamlets. This required village and provincial officials to relearn new regulations and operational procedures after years of being marginalized by hamlet focused pacification programs. The new program provided intensive technical assistance and training on the village, hamlet, and provincial level with a goal of shifting oversight of provincial and municipal development to local administrations by 1970 (USAID 1975b: Vol. I, 34/291, 41/298–42/299).

The Phoenix program

The Phoenix program, also known as Phung Hoang, a top priority after the 1968 Tet Offensive became CORDS' most controversial initiative. It built upon an earlier CIA Intelligence Coordination and Exploitation Program (ICEX) that had been designed as a clearing house for information on the VC. The Phoenix program was focused on exploiting the strengths and vulnerabilities in the VC networks revealed when they had assisted the North Vietnamese Army with their Tet Offensive. Along with their South Vietnamese counterpart, CPDC, CORDS made Phoenix into the centerpiece of "anti-infrastructure operations" for the 1968 Accelerated Pacification Campaign. The original ICEX structure was decentralized, responsibility transferred to provinces and districts where intelligence gathering and operations coordinating centers would be built to interrogate suspects (Andrade and Willbanks 2006: 30, 18). They created files and dossiers to "neutralize" VC by capturing, repatriating or killing suspects and developed judicial rules for trials and imprisonment of suspected VC. Local militias known as Regional Forces or Popular Forces, the National Police, and CIA recruited Provincial Reconnaissance Units carried out these operations. Phoenix advisors were assigned to all 44 provinces, with most districts reaching a total of 704 advisors by 1970 (Andrade and Willbanks 2006: 32).

The decentralization strategy resulted in a wide range of outcomes. In an attempt to unify performance standards the Phoenix program set quotas for neutralization that inadvertently led to fabricated numbers of detainees, false arrests and increased bribes by the VC that resulted in releases of up to 60 percent of prisoners in some areas. Critics accused the Phoenix program of being a rogue operation specializing in circumventing the rule of law, carrying out assassinations, and contracting out their "dirty work." Others pointed out that the program operated under a type of customary law known as An Tri that required three independent sources of evidence for conviction and that two-thirds of their suspects were captured not killed. While many acknowledged that abuses had occurred they also made the point that the VC insurgents were legitimate targets. In retrospect it has been generally acknowledged that setting a quota system that resulted in the support of corrupt practices and false reporting of neutralization numbers may not have been a good public strategy (Andrade and Willbanks 2006: 9–10).

Final revision of roles

The final restructuring of civil–military roles began with the 1968 announcement that US combat forces were to be gradually withdrawn from Vietnam (USAID 1975b: Vol. I, 38/295). This would be accomplished by an accelerated process of "Vietnamization" that replaced American combat forces with Vietnamese in operational and supporting roles. The implications of this transition became clear as the US Ninth Division stationed in the northern Delta of South Vietnam turned over military responsibilities and bases to Vietnamese forces in preparation for leaving in August 1969. US advisory activities were significantly cut back, integrated into Vietnamese government programs, or taken over by Vietnamese who were US government employees (USAID 1975b: Vol. I, 39/296). USAID contributed the majority of personnel and program funding to CORDS as US troops withdrew. Earlier military "pacification" efforts that included intelligence advice, training and expanding Vietnamese forces were restructured. Instead USAID Vietnam assigned CORDS field advisors to programs that focused on refugees, public safety (training of the National Police), provincial and village-based agricultural, educational, public works programs, assistance to the City of Saigon and a successful USAID VC repatriation program called Chieu Hoi or Open Arms (USAID 1975b: Vol. I, 39/296).

By 1971 military pacification and civilian development programs appeared to have made significant inroads into VC operations. Violence and political intimidation began to disappear or subside to acceptable levels and the GVN began to dominate rural areas. Many villages were under the protection of National Police, development programs were effectively managed by local governments, and it appeared that a shift from military operations to a civilian development focus would soon be possible (USAID 1975b: Vol. I, 45/302). Overall CORDS "unity of effort" appeared to have had a significant positive impact on the insurgency with almost 93 percent of South Vietnamese living in relatively secure villages by 1970, up from 20 percent in 1968 (Andrade and Willbanks 2006: 17).

Hanoi's Easter offensive in April 1972, the Paris Accords in January 1973, and declining levels of aid caused by US troop withdrawals and global inflation shaped the final years of US Vietnam policies 1973–1975 (USAID 1975b: Vol. I, 7/22). CORDS disbanded after the ceasefire of 27 January 1973 and all remaining US military forces rapidly withdrew. The US Ambassador set up a new office under a Special Assistant for Field Operations to monitor pacification and development activities with greatly reduced USAID staff (159). USAID continued to support pacification and development planning and program implementation to the extent allowed by continuing violence (USAID 1975b: Vol. I, 52/309, 2/6).

Years later the North Vietnamese confirmed that while the Phoenix program did slow their progress between 1968 and 1970 they learned their lessons well (Andrade and Willbanks 2006: 35–6, 43). Despite a solid military defeat in the 1968 Tet Offensive the North Vietnamese Army won a political victory that set the stage for their 1972 Easter offensive. The final 1975 defeat of South Vietnam by the North Vietnamese Army that resulted in the emergency evacuation of all US personnel brought the US civil–military "unity of effort" to a sudden and final end (Andrade and Willbanks 2006: 9–10).

In retrospect it is hard to imagine the degree of political and social turmoil that consumed the US during the Vietnam War. Unlike later conflicts in Afghanistan and Iraq the American public was fully engaged in Vietnam and almost no one was untouched. It has been called a "profoundly negative experience for the American people and the most divisive event since the Civil War" (Capps 1991: 1). Set against a backdrop of US political and social unrest the sheer numbers of 536,100 soldiers drafted to serve in Vietnam boggles the post-Afghanistan and Iraq mind. American military deaths totaled 58,220 in addition to the numbers of American civilians killed or captured, cost over $7,551 billion, and an additional $343 million per year in US economic aid (National Archives 2008; History Place 1999).

Similar to World War II the Vietnam War shaped the national and world view of an entire generation. The final defeat of a South Vietnamese government that had cost the US so much in blood and treasure triggered a type of amnesia and disassociation among US military forces and civilian organizations. The war was so uncomfortable for the military that the US Army "consciously decided to forget" by locking away the records and pretending "that nothing had ever happened" (Stout 2013). Anticipating similar reactions among USAID and other government agencies the 1975 Viet Nam Terminal Report stated it was a "factual account of the times" written to "preserve primary source material of historical value for the 20 years of USAID work in Vietnam" (USAID 1975b: Vol. I, Summary, 3).

Lessons to learn from Vietnam

US experience in Vietnam marked an equally important and contested turning point for civilian and military roles. CORDS did prove that civil–military organizations and programs could be coordinated effectively under a single manager in combat conditions (Andrade and Willbanks 2006: 16–17). While the US military contributed the majority of personnel, funding, and resources, civilians held significant policy-making positions and served as field advisors alongside military forces, which improved civil–military cooperation. Communication between CORDS staff and the South Vietnamese government became clearer and helped improve South Vietnamese pacification programs especially ones that supported local militias and encouraged them to challenge contested areas and fight the VC's political–military infrastructure (Jones 2005: 115–16). However, often missed in contemporary interpretations of CORDS is the extent to which the transition from separate but equal civilian and military efforts to a CORDS model merging development and security in combat conditions was internally and publicly questioned at the time. The Phoenix program was especially controversial and continues to raise important questions now about how to separate and determine the most effective roles for civilians and military forces in dangerous environments. Two key lessons from Vietnam applicable to future interventions beyond Afghanistan and Iraq are that core US government and military goals need to be clearly articulated to develop civil–military programs that can effectively respond to changing environments, and that civilian and military roles need to be clearly and carefully defined and redefined in dangerous environments.

Setting clearly articulated and adaptable goals

While the US did pursue defined objectives in Vietnam it kept changing its mind as to what they were. The reason for initial US involvement was to stop the spread of communist ideology into South Vietnam. This was reinforced by the popular domino theory that one country succumbing to communism would lead to the downfall of an entire region and early USAID programs and US military advisors focused on strategies that would hold that line. Civil–military programs were not prepared or equipped to keep pace with the tempo of violence that grew from a relatively low key VC guerilla (word used to describe terrorists, insurgents, revolutionaries) insurgency to a full scale conventional war. The CORDS "unity of effort" model was arguably a last ditch effort initiated late in the war past the point when it might have made a real difference.

Years later several US officials conceded that the domino threat had been disproved when Indonesia successfully underwent an anti-communist revolution in 1965, the same year that President Johnson began the major US military commitment in Vietnam (Fromkin and Chace 1991: 96). By the middle of the 1960s a number of leaders in President Johnson's administration began to feel that the US was fighting a war to impose a regime that even it thought was unsatisfactory on "a country of no clear importance" to the US (Fromkin and Chace 1991: 92). In May 1967 Secretary of Defense Robert McNamara wrote in a memo that said he no longer "believed it vital that South Vietnam should remain independent or that it should remain non-communist," words he later confirmed indicated he no longer saw a compelling reason to continue fighting (Fromkin and Chace 1991: 92). When Clark Clifford who succeeded McNamara as Secretary of Defense consulted with the Joint Chiefs of Staff it became clear to him that the entire US strategy was to continue pressure until it became "unbearable for the enemy" who would be forced to capitulate. Unfortunately no one knew if the bombing campaign in North Vietnam was effective, how many more soldiers would be needed to "win" or how to iden-tify when that point would be reached (Clifford 1991, 148).

The evolution of civil–military programs and roles throughout US involvement in Vietnam reflected this uncertainty. They were constantly changed in attempts to "fix" or respond to problems that no one in US government seemed to understand or be able to define. The only universal agreement about reasons for US involvement was that they were not fight-ing to make South Vietnam into an American colony. Ironically that was precisely what many Vietnamese believed to be the case (Fronkin and Chace 1991: 91).

A poll taken in 1975 showed that 70 percent of US Army generals believed that it was not clear what America had hoped to achieve in the Vietnam War. The key lesson for 91 percent was that if the US were to ever fight this type of war again it should begin by deciding what it wanted to accomplish (Fronkin and Chace 1991: 91). Lyndon Johnson's "other war" of pacification could not transform a profoundly corrupt politically unstable government saddled with a lack of political will and ability to effectively execute plans and pro-grams (Jones 2005: 116). Despite a number of potentially positive outcomes historian Richard A. Hunt later described the achievements of CORDS and pacification as "ambiguous" at best (Hunt quoted in Andrade and Willbanks 2006: 21–2).

Defining civilian and military roles in dangerous environments

It is clear that many USAID civilian personnel were frustrated and con-cerned as their mission abruptly changed from a separate humanitarian aid and development effort to full participation with the military in the midst of combat conditions. These included strategic hamlet programs that moved villagers into guarded camps and programs that operated in areas after saturation bombing and search and destroy tactics had countermanded other positive civil–military efforts (Dixon 2009: 363). Reverberations from CORDS programs connected to intelligence agencies blurred civil–military roles and created longstanding questions about distinctions between pro-grams that gather intelligence and civilian-led projects. The legacy from the Phoenix program can be argued as the most problematic, with many unin-tended consequences for future civil–military collaborations. These activities have caused distrust about the true intent of US civilian advisors, humani-tarian workers and researchers carrying out aid programs in other countries. There are still suspicions that US civilian organizations are carrying out CIA intelligence activities under the guise of aid and development programs that are reinforced whenever a civilian program is used by intelligence organiza-tions as a cover for covert activity. This doubt has contributed to the target-ing of foreign and national civilian humanitarian and aid workers and makes well intentioned civil–military coordination more problematic than it might otherwise be.

In 1970 CORDS director Komer commented that while the "new model" pacification program was "fairly good," "other things" were very bad so while there was a chance that the US might benefit by comparison he had "the grim feeling that we may end up with a disaster in Vietnam" that would eclipse all positive efforts. At that point no one except possibly "professional historians" would ever "be able to delve in and find out that maybe the whole thing collapsed, but the pacifiers at least were on a reasonably promising track" (Komer 1970: 11–12). USAID's terminal report echoed these sentiments stating that "in view of the collapse of the GVN" it was difficult "to evaluate any US assistance activities in South Vietnam as having been successful, "yet much good was accomplished" in many areas, especially the agricultural sector, "which doubtless will endure regardless of the new form of political organization now emerging in the country" (USAID 1975c: Vol. II, 25).

In a 1965 essay Hobsbawn noted that since World War II the US had bet on its superior strength as an industrial power and "its capacity to throw ... more machinery and more explosives" into a war than anyone else but came to face its limitations in a "guerilla" war in Vietnam that undermined conventional military operations (Hobsbawn 1998: 200–1). Observing that while the "guerillas of today" might have better equipment than their predecessors their true strength came from their political appeal to the common interest of poor against the rich, the oppressed against the government, and through exploiting nationalism and "hatred of foreign occupiers" (Hobsbawn 1998: 202). This was misunderstood by foreign forces who thought that the balance could "be tipped to achieve success by one more effort, more troops, more bombs ... more social missions." Hobsbawn maintained that the French experience in the 1950s Algerian War and later attempts to hold onto their Vietnamese colony "foretold the American story." Despite deploying approximately one soldier to 18 nationals in Algeria, France "could only hang on there for seven years" and their "occupation of Vietnam lasted for nine years before they were forced to leave" (Hobsbawn 1998: 207, 209). As the international community struggles to cope with future scenarios an understanding of how unlearned lessons in Vietnam shaped US civil–military roles and expectations in Afghanistan and Iraq may help make international efforts more effective in the future.

2 "Unity of effort" in the long wars

Large scale integrated civil–military strategies that evolved in the years following Vietnam were attributed to the United Nations' experiences in Bosnia and Kosovo during and following the Yugoslav Wars, 1991–1999. During these complex peacekeeping missions economic assistance and government infrastructure was established that developed good governance and accountable legal systems. Military forces were called upon to escort humanitarian relief supplies and participate in civil–military cooperation operations (CIMICs). As the number of non-governmental organizations (NGOs) and independent civilian organizations in conflict areas increased so did contact between civilian organizations and military forces operating in the same areas (Van de Kuijt 2012: 20–1).

Humanitarian aid was still assumed to be the primary responsibility of governments and aid agencies who agreed to abide by basic principles of "humanity, impartiality, and independence" regardless of political affiliation to save lives and relieve suffering. They provided food, shelter, water, sanitation, and emergency health services to populations affected by armed conflicts and natural disasters. Development assistance was also considered to be the responsibility of governments and large international NGOs who provided financial and material resources to reduce poverty, promote economic and social development and support the wellbeing of populations in developing countries. While international development funding was often linked to overtly political goals, such as supporting specific political transformations, it was widely understood that donors should refrain from directly linking aid to a pursuit of their own political objectives (Oxfam International 2011: 7–8).

Prior to 2001 many international humanitarian and development organizational conversations focused on improving aid effectiveness in alleviating poverty and conflict. Discussions about international roles for military forces among international civilian organizations were often linked to the concept of the responsibility to protect also known as R2P that defined the role of security forces in protecting vulnerable populations during peacekeeping missions, peace processes, and agreements (see Chapter 4). There was little conversation about working alongside military forces other than how to utilize their logistical support for fast tracking emergency supplies and aid to international civilian organizations responding to disasters.

These conversations changed following the US-led invasions of Afghanistan and Iraq as civil–military roles and operational relationships evolved along two similar but separate paths in each country. The Afghanistan War, which had begun on 7 October 2001, was an internationally endorsed response to the 11 September 2001 attacks, with a clear mission to find and defeat the Taliban who had been harboring Al Quaeda operatives. The invasion of Iraq on 20 March 2003 was much more internationally divisive and perceived by many to be an unprovoked US military operation. While invasions of both countries were led by the US military there was a greater degree of international civilian and military engagement in Afghanistan than the primarily US focused interagency effort in Iraq.

As both wars evolved the US-led "unity of effort" and "hearts and minds" civil–military missions increasingly drew stabilization and reconstruction policies from Vietnam era models described as a "whole of government" approach that included pacification, reconstruction and peace–support operations in each country. This was followed by a "hearts and minds" counterinsurgency (COIN) strategy that assumed, similar to Vietnam, that violence driven by government neglect and poverty could be stopped by civil–military programs to improve governance and increase economic opportunities in each country (Jackson and Haysom 2013: 9). Short term goals focused on winning the public's approval for their government while laying the groundwork for long term political, social, and economic changes. This concept was problematic for many international humanitarian and aid organizations who perceived these aid strategies to be driven more by political and military objectives than by local needs. Many were also concerned that this highly integrated civil–military approach would increase the vulnerability of their civilian staff and the populations they were serving (Jackson and Haysom 2013: 9–10).

This chapter gives a general overview of each war rather than a comprehensive history of each conflict, which is recent and well documented to provide context for tracing the evolution of contemporary civil–military programs and roles. It explores the emerging concept of civil–military coherence and discusses how civil–military experiences in both Afghanistan and Iraq set the stage for examining future roles.

Afghanistan, 2002–2014

Post-invasion Afghanistan was a fragmented unstable country with endemic poverty and a divided society that had survived decades of internal wars and external interventions (CDA Listening Project 2009: 5). The international aid community had a long history of humanitarian engagement there. Several organizations continued their operations under Taliban control, human rights organizations had frequently publicized Taliban abuses toward women and girls, and aid organizations had worked with millions of Afghans in Pakistan's refugee camps. Following the Taliban government's collapse in December 2001, senior elite Afghans met in Bonn, Germany, to form a 6 month interim government that would be succeeded by a 2 year term Transitional Authority. The United Nations formed the International Security Assistance Force (ISAF) with the first troops deployed by June 2002. Oversight was transferred from UN to NATO (North Atlantic Treaty Organization) control over 9,700 troops by August 2003 (Jackson and Haysom 2013: 9).

The war in Afghanistan can be organized into three phases. The first was a phase of stabilization and reconstruction from 2002–2008, a "surge" of civil–military personnel took place from 2009–2011, with a drawdown and transition to Afghan government leadership that began in 2012, with cessation of combat operations at the end of 2014 (Jackson and Haysom 2013).

2002–2008

Four US civil–military Provincial Reconstruction Teams (PRTs) were established in 2003 to stabilize their areas of operation and act on behalf of the Afghan government until it could provide services and security to outlying regions. These were modeled after the earlier US-led Coalition Humanitarian Liaison Cells, which had been made up of 10–12 military forces assigned to gather intelligence and implement quick impact projects. These became Joint Regional Teams, then Provisional Regional Teams and finally PRTs, which had increased to 19 by 2004. ISAF declared them to be civil–military institutions but in reality they were staffed with 50–100 military members and few if any civilians. Designed as cost effective alternatives to large numbers of troops and financial resources PRTs were tasked with establishing small zones of stability by identifying potential projects and civilian partners that included NGOs, UN affiliates, and the Afghan Transitional Authority that could be linked until an entire region was secure. The small number of international aid organizations who had been allowed to operate under Taliban leadership saw their numbers rise to over 800 aid agencies by 2006. However funding was still relatively limited with less than 10 percent of what had been previously allocated to Bosnia after the Yugoslav War and less than 25 percent of that given to East Timor (Jackson and Haysom 2013: 9–10). While many humanitarian and aid organizations were wary of the military component of PRTs the relatively low numbers of ISAF security forces (16,700 by 2004) made participating with PRTs necessary to ensure the security of many NGOs and their staff who requested that they be expanded throughout the country (Jackson and Haysom 2013: 10).

Concerned, aid practitioners with Afghanistan experience discussed ways to develop a more accurate system for evaluating the impact of international programs on ordinary Afghans in a 2003 London workshop. Participants emphasized that history mattered when trying to understand the post-2001 Afghan environment. Prior to that time outside assistance had consisted of short term development projects with defined expectations, clear boundaries and no links to security. The presence of ISAF was significantly changing the conditions linked to these programs and having a negative effect on the security of international civilian workers. By 2003 the influx of $300 million US aid had fuelled a boom economy of $2 billion, which when combined with unreported remittances from Afghans working abroad, diaspora communities, and income from black markets had created a freewheeling unregulated economy. This increased uneven distribution of wealth and benefits among Afghans and made conventional economic measures unreliable indicators in measuring the wellbeing of the overall population. Participants listed key issues that included better identifying when and in what circumstances security forces were helping or hindering aid and humanitarian assistance, how to protect aid workers and recipient populations with or without the input of Afghans, and having the Afghan population prioritize internal security threats. They recommended having Afghans identify vulnerable populations and those who were internally displaced to create more accurate numbers of internally displaced populations that would lead to more accurate indicators to assess human welfare including food, sanitation and health infrastructure (Workshop for Post-Conflict Afghanistan 2003).

ISAF had assumed supervision of all PRTs by 2006 but their activities were limited by the constraints of the local security environments and the policies of each country supplying military forces and civilians. PRT sizes and organizational structure varied widely but civilians generally made up about 5–10 percent of any staff. US PRTs had ratios of about 100 military staff to five civilians while German PRTs had up to 300 military forces for a small number of civilians. There were also significant differences in leadership structures. US PRTs were always led by the military, the UK's were led by a civilian, and Germany's were split between a civilian and military leader. The end result was chaotic, with little coordination or structure and no clear goals or mission (Jackson and Haysom 2013: 10–11).

The US Agency for International Development (USAID) underwent three different phases between 2001 and 2007. Beginning small their 2002–2003 focus on humanitarian and short term assistance to displaced persons and food programs to avoid famine was soon eclipsed by quick impact road construction and infrastructure projects. From 2004 to 2006 USAID developed a comprehensive program that focused on addressing a wider range of needs that included agriculture, education, health and generating power. The US General Accounting Office observed that it was not clear whether USAID's reconstruction programs had led to improvements or whether its goals were too widespread to achieve significant results. It also reported complaints that Kabul and Kandahar provinces where security concerns were highest had received 70 percent of allocated funds for road construction along with targeted efforts to provide economic alternatives to areas where opium poppies were cultivated. As none of these projects had been evaluated it was difficult to judge whether they had achieved their desired results (GAO 2007: 30–1).

Civil–military relationships began deteriorating after the UN Office for the Coordination of Humanitarian Affairs (OCHA), operating in Afghanistan since 1988, closed in 2003 and was incorporated into the newly formed United Nations Assistance Mission in Afghanistan (UNAMA). Only one officer was assigned to civil–military affairs for all of Afghanistan compared to ISAF's 35 CIMIC officers in Kabul headquarters (Metcalfe *et al.* 2012: 1–3). There was also a range of interpretations for civil–military roles. OCHA and the Inter-Agency Standing Committee defined civil–military coordination, also known as CMCoord, CIMIC, and civil–military relations, as an essential dialogue and interaction between civilian and military actors in humanitarian emergencies that ranged from co-existence to full cooperation necessary to protect and promote humanitarian principles, avoid competition, minimize inconsistency, and when appropriate to pursue common goals. CMCoord referred to specific civil–military interactions for humanitarian purposes while civil–military relations referred to a broader range of interactions that included civil society, governments, legal, human rights and development sectors. CIMIC, a military term defining coordination and cooperation with civilian populations, local authorities, international and national NGOs and agencies in support of a defined mission, sought to establish links between civilian agencies active in a military force's theater of operations. Their purpose was to obtain support and possibly assets for military priorities but did not guarantee a humanitarian response with conditions defined by civilian organizations. CIMIC officers belonging to the UN Department of Peacekeeping Operations (DPKO) defined their role as facilitating and preserving civil–military relationships, interactions, and dialogue that included civilian humanitarian and development organizations operating within the mission area in support of UN Security Council and mission objectives. European Union CIMIC objectives were similar to the UN's (Metcalfe *et al.* 2012: 1–3).

UNAMA attempted to coordinate and clearly define when and where PRT interventions would be most effective to avoid duplication of PRT and aid agency activities. This resulted in the Principles Guiding PRT Working Relations with UNAMA, NGOs and Local Governments in 2003 but had little effect on the military's quick impact projects. A 2004 UN working group was poorly attended by NGOs and the military representatives were continually changing. A 2004 PRT Executive Steering Committee issued guidance for PRT policy and external interactions, but lack of ISAF coordination of PRTs meant that information was not communicated to the relevant leaders. A final significant initiative during this pre-surge phase was formation of the Civil–Military Working Group in Kabul and throughout Afghanistan, which included NGO, UN, and military representatives. Their goal was to facilitate communication and coordination between humanitarian actors, international military forces and other "stakeholders" in order to identify and address mutual areas of concern. They created the Civil Military Guidelines, which reinforced international standards and practices adapted for operations in Afghanistan that included defining roles for civilians and military in humanitarian operations. Endorsed by UNAMA, ISAF, and the Agency Coordinating Body for Afghan Relief these were widely circulated throughout the civil–military communities in 2008. While viewed as important to the international civil–military consensus building process there was little or no input by Afghan NGOs and they had limited practical application in an environment where the resurgence of the Taliban was rapidly deteriorating security. Deaths of Afghan civilians increased 60 percent in 2006 and international aid workers were increasingly targeted. By 2008 nearly half the districts in the country were considered too dangerous for UN civilians traveling on their own (Jackson and Haysom 2013: 14, 16).

2009–2011

By 2009 the US, European Union, United Kingdom, Japan, and over 60 countries and international financial institutions had contributed to reconstruction and development effort and 46 countries had provided troops to ISAF. Facing an increasingly violent environment the US and international allies decided to dramatically increase military forces shifting from a comprehensive integrated civil–military approach to a military-led "hearts and minds" COIN strategy in 2009. Military forces had nearly doubled from 56,000 to 132,000 by the middle of 2011. They dominated civil–military roles and relationships while civilian aid organizations became increasingly concerned about the blurring of humanitarian roles. Following the pattern of COIN operations in Iraq (see below) and earlier in Vietnam (see Chapter 1) US and international military forces focused on clearing and holding dangerous areas so that civilian agencies could implement projects. Sustainability was rarely considered in the rush to secure areas and deliver what military members called "government in a box." By 2010 the numbers of Afghan civilians who were assassinated by the Taliban for associating with government officials and/or working with military forces more than doubled as did attacks on military constructed projects and schools (Jackson and Haysom 2013: 17, 20).

This was a particularly difficult period for civil–military relationships in Afghanistan. ISAF was no longer viewed as a peacekeeping force with increasing military operations. Many international civilian aid agencies began distancing themselves from military forces and UNAMA who they felt that despite statements against pressuring humanitarian organizations to support military efforts in 2010 had not sufficiently advocated for separation of roles. Conversely, increased military activity and adoption of a COIN strategy opened up an opportunity for engaging civil–military groups in a new conversation focused on protecting the civilian population. As a result ISAF instituted new Tactical Directives that restricted the use of force against civilians that significantly reduced deaths by 2012. Ironically, as numbers of ISAF caused deaths reduced the numbers of civilians killed by the Taliban rose (Jackson and Haysom 2013: 17, 20–1).

2011–2014

The ISAF troop surge ended in September 2012 with forces gradually reduced to pre-surge 2009 levels. As the insurgency continued to spread and the Afghan government continued to underperform, the international community, determined to leave Afghanistan, developed two strategies designed to achieve an acceptable level of stability that would allow an "honorable exit." The first was a phased transition to withdraw combat troops by the end of 2014 while continuing financial support for Afghanistan's civilian and military institutions. The second was a process of reconciliation structured to negotiate a settlement with insurgents implying a change to the Afghan constitution. Both projects had contradictory goals, with the first assuming the state would be strong enough to maintain constitutional order and resist an insurgency. The second assumed that the insurgents would not be defeated and that political structures would need to be re-negotiated with significant constitutional amendments to accommodate them. Simultaneous negotiations resulted in failure for both before the 2014 presidential elections with a fumbled opening of a Taliban political office in Qatar in June 2013 and the refusal of President Karzai to sign an agreement allowing a residual international training force and setting conditions for financial assistance to Afghan army and police. The only option that remained to resolve the country's crisis was the 2014 presidential election that would allow Afghan political elites to extend political processes with and achieve a level of legitimacy that would allow them to pursue either or both tracks (Smith 2014: 2–3).

As the 2014 withdrawal date approached, military forces prepared to hand off security operations to the Afghan National Security Forces and PRTs to Afghan and international humanitarian and aid operations. Donor contributions to humanitarian and aid projects were to be significantly reduced, power relationships realigned, jobs eliminated and vulnerability of Afghans who worked for ISAF increased (Jackson and Haysom 2013: 23). Afghanistan held two rounds of presidential elections beginning 4 April 2014 that resulted in the election of Ashraf Ghani in a confused contest of voting and narratives. During this process the Afghan media stopped reporting Taliban attacks, both leading candidates minimized allegations of fraud after the first round of elections, and the Taliban decided to engage in the electoral process after what was perceived to be a propaganda defeat (Giustozzi and Mangal 2014: 5). At the end of 2014 combat operations had officially ended but violence increased as a large residual force of up to 12,000 NATO troops remained behind to train, advise, and assist Afghan forces. An estimated 10,000 of these forces were supplied by the US military (Donati 2014). As international civil–military communities continued to reconfigure their operations the Afghans were once again left to cope with another fundamental national transition (Jackson and Haysom 2013: 23).

Iraq, 2003–2011

Iraq had been greatly weakened by UN sanctions when US forces invaded Iraq on 21 March 2003; however, they still had a strong governance and physical infrastructure intact. Launched with a minimum of international support other than the United Kingdom, the invasion was a military success, and while military forces could provide short term security it soon became clear they would not be sufficient to establish long term security, stability and a reformed government (Bucar-Marcu and Fluri and Tagarev 2009). As a result, a US civilian-led Coalition Provisional Authority (CPA) was created with full legal and executive authority over Iraqis in April 2003. Before authority was returned to the Government of Iraq in June 2004 the CPA enacted a number of laws seen as controversial by Iraqis and the US civil–military community. These included disbanding the Iraqi Army, which was widely credited with starting the Sunni insurgency and giving immunity to private security contractors operating in Iraq including Blackwater who were later accused of the unprovoked killing of Iraqi civilians (Tristam 2014). CPA actions taken during this phase were considered by many in the civil–military community to be directly linked to the violence that followed. One of the chorus of civilian and military voices later asking why post-invasion Iraq had gone so badly, Stephen Van Evera stated that unlike World War II when the Marshall Plan sought to employ staff of the highest caliber "lesser standards were applied" to appointees leading the Iraq CPA that resulted in "incompetents" being appointed and "inferior results" achieved (Van Evera 2007: 32, 64).

The United Nations Assistance Mission for Iraq (UNAMI) was formed by Security Council resolution in August 2003 to function as the umbrella organization for a "cluster approach" or "family of UN Organisations involved in Iraq" known as the United Nations Country Team. The 16 member UN agencies and programs included the Office of the UN High Commissioner for Human Rights, United Nations High Commissioner for Refugees (UNHCR), and the United Nations Development Programme (UNDP). There were two "affiliated bodies," the International Organisation on Migration and the World Bank. It was internally organized by agriculture, food security, environment and natural resources management, education and culture, governance and human development, health and nutrition, infrastructure rehabilitation, refugees, internally displaced persons and durable solutions, and support to the electoral process (UNAMI 2007; 3; UNAMI 2008).

The UN headquarters in Baghdad was destroyed by a suicide bomber on 19 August 2003 that resulted in 22 deaths and severely injured survivors. The UN Secretary General Kofi Annan withdrew all staff from Iraq and reposted them in Jordan and Kuwait. A core group returned to Baghdad in April 2004, and when the security environment improved staff returned to Baghdad in increasing numbers during the spring of 2009. While UNAMI's engagement with US military forces was perceived by some to have been the cause of the attack it is possible the UN underestimated Iraqi resentment toward them for imposing sanctions and for a perception that the UN had not protected vulnerable groups. A 2004 survey of Iraqis revealed that many felt UN sanctions had undeservingly punished portions of the population who had no control over Saddam Hussein's actions and that the UN had stood by when Hussein had punished the Marsh Arabs in the South and the Kurds in the North in the 1990s for rising up against his regime (Stoddard *et al.* 2009: 8; Iraqi Voices, 2004).

2004–2005

The inability of the US military and new Iraqi government and security organizations to restore order to the country began undermining its legitimacy by 2004. With no standing Iraqi Army remaining the looting that began in April 2003 disintegrated into criminal networks, insurgent groups and militias, and a bloody sectarian civil war (Simon 2007: 15). As violence increased the US Army and others renewed calls for a Vietnam CORDs-like "unity of effort" political, economic, and security counterinsurgency approach that would establish an integrated civil–military organization that combined US civilian organizations and the Multi-National Force-Iraq (MNF-I) under one "unity of command" (Coffey 2006: 24).

The State Department established civilian-led PRTs in 2005 as "an important tool in achieving stability in Iraq" with a mission of "bolstering moderates, promoting reconciliation, fostering economic development and building provincial capacity" (PRT Fact Sheet 2009). Staffed by a range of US civil–military personnel they included employees of the Department of State, USAID, Department of Defense civil affairs, US Department of Justice, US Department of Agriculture, US Army Corps of Engineers, contracted Subject matter experts and local personnel. Their mission was to promote stability and development at the provincial level, while focusing on governance, economic development, national unity, political development, and the rule of law (PRT Fact Sheet 2009).

Based on their recent experience in Afghanistan and Kosovo the international humanitarian community warned about the danger of creating too little humanitarian space in Iraq. A 2003 briefing by the UK Humanitarian Policy Group warned that the combination of military action, Iraqi dependence on imported food, and breakdown of the UN Oil for Food Programme meant there were likely to be high levels of vulnerability among Iraqis who might not receive any of the large amounts of US assistance being committed unless an assessment was taken to identify those who needed help the most. Unlike Afghanistan where limited numbers of aid organizations had been continually operating with a small but viable network there was a very small international presence in Iraq and very few Iraqi civilian aid networks to call upon to assist in meeting internal humanitarian needs. As the US and UK governments were partners in the invasion and major funders of humanitarian needs many NGOs as well as the UN were concerned about perceptions among the Iraq population that they were operating as part of their country's military and foreign policy agendas (HPG Briefing Note 2003).

The Humanitarian Policy Group identified three specific areas of concern in Iraq. One was the goal of "hearts and minds" operations where the emphasis on security objectives could potentially override humanitarian operational goals to save and protect lives and prevent suffering of all populations. There was also a tendency for military-led humanitarian efforts to favor particular groups tied to their location and perceived current or potential assistance and support. From a civilian humanitarian perspective this strategy could easily bypass individuals and groups who most needed assistance. The extensive military provision of humanitarian assistance made it difficult for other international and local civilian aid staff to separate themselves from the military operations, thus losing their "perceived neutrality" and making them more likely to become targeted by non-state armed groups. A third point challenged the effectiveness and efficiency of the military providing humanitarian assistance, as UN guidelines stated that humanitarian work should be performed by humanitarian organizations and in cases where military organizations had a role there should be a clear distinction made between the functions and roles of humanitarian organizations and military forces (HPG Briefing Note 2003).

2006–2008

Violence escalated in 2006 as Iraqis living south of Kurdistan found themselves under siege. The bombing of the Shia al-Askariya Shrine mosque and shrine in Samara on 22 February 2006 unleashed a firestorm of "unprecedented" sectarian violence that displaced hundreds of thousands throughout Iraq (Ajami 2007: xix). While violence against Shias had increased in the previous 3 years this attack on one of their great religious monuments located in the midst of a predominantly Sunni town provoked both Sunnis and Shias into an unprecedented sectarian war. Fouad Ajami wrote that "the enemies … had proven unusually brutal … and unusually skilled in the war they had waged against the new order of things in Iraq" (Ajami 2007: Introduction). By March improvised explosive devices and vehicle-borne improvised explosive devices were being discovered daily and gangs of armed men were roaming Baghdad streets.

The Iraq government's inability to impose law and order after its army was disbanded undermined its legitimacy and relevance adding to the fear and resentment driving insurgent violence. Exuberant celebrations and lawless looting that had begun with regime change in April 2003 rapidly degenerated into a crime wave then an insurgency and violent sectarian fighting. The inability of the US armed forces and the Iraqi security services to control the south and center of the country allowed criminal gangs, insurgent groups, and militias to continue their violence. What began as resistance against the US occupation evolved to fighting on religious grounds as armed groups took control of major towns and cities. As violence peaked several US, international and Iraqi NGOs were recognized for reaching across sectarian lines to "enhance dialogue and understanding" despite the dangers, but the majority of international humanitarian and aid organizations were constrained by the lack of security. A highly politicized government process restricted Iraqis who did have access to specific areas from forming and registering their own NGOs (Baker and Hamilton 2006: 15, 17, 26).

It became clear that without a dramatic increase in military forces the US would have to withdraw. In 2007 a decision was made to "surge" US forces by adding four more army brigade combat teams with up to 4,000 soldiers each to the approximately 138,000 troops they had since 2004. Military numbers began overwhelming civilian counterparts. At that time USAID had a total of 2,000 officers deployed worldwide while the State Department had 6,000 Foreign Service Officers whose primary function was to serve as diplomats in 270 embassies, consulates, and international posts. Other US agencies were in a similar situation (Davidson 2010: 167). Seven PRTs embedded with military brigades following the 2007 surge and by mid-December 2007 there were 28 US PRTs staffed by approximately 700 people located in all 18 of Iraq's governorates, with an additional 15 embedded with military forces. There were also PRTs from other MNF-I troop donor countries including Italy, Great Britain and Australia (Shearer 2008: 6).

Beyond the UK and coalition partners (including Italy, Australia, Romania) the international community continued to play a limited role in Iraq. Despite their small presence the UN helped Iraq to hold elections in 2005, draft the constitution, organize the government and develop institutions. The World Bank committed limited resources with one or two staff based in Iraq while the European Union had one representative. Other than the US the UK had the second largest civil–military presence with 7,200 troops and a strong diplomatic presence in Basra and the southeast. They confirmed their commitment to continue working for stability in Iraq only reducing troop numbers and resources when the situation on the ground appeared to support that action (Baker and Hamilton 2006: 26–7). The US military surge did succeed in stabilizing the security situation with assistance from members of the insurgent Sunni Sons of Iraq militant groups who switched sides to help the US forces fight and successfully repel Al-Qaeda insurgents. This created a more secure environment that allowed a significantly increased civilian presence to engage in the reconstruction efforts.

2009–2011

By 2009 there were thousands of civilians surrounded by approximately 150,000 troops. USAID employees had increased to 8,610 including Foreign Service and Civil Service Officers, with 6,199 deployed overseas by the end of 2010. The State Department had more than 13,000 Foreign Service Officers and drew upon a Civil Service corps of over 10,000 employees to provide continuity and expertise (Department of State—USAID 2010: 5). Despite hiring thousands of private contractors and temporarily reassigning and appointing government employees significant capacity and funding gaps between US military forces and civilian organizations remained. This discrepancy continued as US civilian and military teams prepared for the final transition from military to civilian leadership for military withdrawal in 2011. Capability gaps between civilian organizations and military forces arose from differences in employee training, focus, and organizational goals. The difference in budget allocations—the Department of Defense budget (FY 2012 request for $670.9 billion) had always dwarfed the State's (FY 2012 total approved for Department of State/USAID budget was $47 billion, 1 percent more than FY 2010 levels) and other interagency partners further limited the scope of military to civilian transfers (Department of Defense 2011: Bureau of Resource Management 2011).

Following a year of negotiations the Joint Campaign Plan outlining the transition from military to civilian leadership was agreed upon in 2009. Many incoming US Forces-Iraq and US Embassy-Baghdad personnel had the cross-organizational experience necessary to realistically discuss capacity, capability, and budget issues. This resulted in a relatively cooperative and flexible process from military to civilian transition. Programs were further reduced by unexpected congressional cuts in the State Department's budget but the State Department's leadership in Iraq at the end of 2011 would still be their largest overseas operation since World War II (Sheridan and Zak 2011).

An agreement was made in early 2009 between the Deputy Special Representative of the Secretary General of the United Nations and the Deputy Commanding General of the Multinational Force in Iraq to share information on humanitarian, development, and reconstruction activities throughout Iraq as part of the "responsible withdrawal" of US forces. Previous information sharing had been based on taking into account "all possible consequences" for the protection of civilians, security of UN personnel, NGOs, and "other relevant stakeholders including any potentially negative effects on beneficiaries/civilians." It now became organized into data sharing groups composed of US military and UNAMI representatives who identified mutual areas of interest and methodologies for exchanging information between both parties (Shearer 2008: 3). This initiative launched a military declassification project for all relevant open source material that had been unnecessarily restricted.

A timetable was established for withdrawal of combat troops by August 2010, with complete withdrawal of all forces by December 2011. By late summer 2009, it was clear that Iraqis had mixed feelings about US withdrawal. While the 2007 military "surge" had helped lay the groundwork for greater security it had not solidified desperately needed political and institutional reforms. There were issues and delays in reintegrating the Sons of Iraq who had been given immunity for their role in assisting US forces during the surge. This was linked to the fact that the original deal had been made between the

US and Sunni insurgents, leaving the Shia Iraq government as a peripheral third party. As the US combat troops withdrew they were replaced by new Advisory and Assistance Brigades formed to "empower" Iraqi Security Forces and civil institutions throughout Iraq. MNF-I Commander General Ray Odierno stated that US forces had a "once-in-a-lifetime opportunity to build a strategic partnership with a Middle Eastern country" (MacLeod 2009).

As the US withdrawal continued and attention shifted to the 2010 parliamentary elections there were chances for political alliances to be made across ethno-sectarian lines in what could become the "transformative moment," which Iraq had not achieved in the first 2005 elections (Visser 2009). UNAMI identified their critical priorities as assisting the preparations for national elections and finding a resolution for Iraq's disputed internal boundary areas between Kurdistan and the remainder of Iraq by establishing the High Level Task Force and the Article 23 Committee in Kirkuk (UNAMI Focus 2009: 2–3). Despite an election in which Iraqis appeared to have voted in a new non-sectarian unity platform old politics prevailed and it was clear that internal challenges were far from over. As a Baghdad street vendor stated prophetically to reporter Anthony Shaddid in 2009 "The flames have disappeared. It's true. … But the war continues among the politicians" (Karrada Abboudi quoted in Shaddid 2009).

Seeking coherence

Based on their experiences in Afghanistan and later observations in Iraq many international humanitarian and aid organizations expressed discomfort with what they perceived as a merging of humanitarian and political agendas that put their operational independence at risk. This resulted in official UN support of integrated approaches to resolving conflicts that called for "greater coherence," clarity, and distinction in civil–military responses (Van de Kuijt 2012: 20–1). The level of coherence achieved in the whole of government efforts in Afghanistan and Iraq varied. Successful efforts required a wide range of defense, diplomacy, development, local and international NGOs and others to work together at some level of cooperation. (Key issues are discussed further in Chapters 3–6.) Each intervening country had a distinctive civil–military cultural approach to synchronizing their response to the different context and phases of each conflict. The whole of government approaches utilized in Afghanistan included the CA or comprehensive approach that added political and socioeconomic goals to a military focus and a 3D approach developed by Canada that combined defense, diplomacy and development (Van de Kuijt 2012: 23, 5). Afghanistan's integrated civil–military PRTs or Provincial Reconstruction Teams were perceived by many to be driven by a primarily political military focus.

Origins of the CA civil–military strategy can be traced back to the beginning of the post-Cold War era when militaries enforcing peace operations shared the same operational space with local populations and international civilian organizations. The European Union (EU), UN, and NATO became increasingly active in delivering and providing aid within this new crisis management framework for multi-dimensional peacekeeping. As a result the CA missions in Afghanistan included a wide range of international political, civilian, military, government and non-government actors (Van de Kuijt 2012: 19).

The 3D concept is an integrated defense, diplomatic, and development model originating with the Canadians who combined their Ministry of Defence, the Ministry of Foreign Affairs, and relevant state development ministry or agencies. The effectiveness of 3D was challenged by a 2007 Canadian briefing paper that asked how long military forces could simultaneously carry out military engagements and development projects in the context of increased insurgency and counterinsurgency effort that resulted in destroying military funded development projects and increased targeting of Afghan, international aid workers and other civilians associated with military forces in the southern provinces. Considered too narrow in scope many ISAF countries including the UK, The Netherlands, and Canada broadened this model during their intervention in Afghanistan to include their national justice, economic affairs, and policing ministries or departments (CCIC-CCCI 2007; Van de Kuijt 2012: 22).

An updated version of the Vietnam era Province Reconstruction Teams (see Chapter 1) provided the basic structure for the PRTs or Provincial Reconstruction Teams revived by the US in Afghanistan in 2002 then developed and utilized throughout post-2003 Iraq. The Afghanistan version combined members of the military with civilians assigned to development projects, programs, political analysis, and police and legal advisors (Van de Kuijt 2012: 22). Their dominant military component also made PRTs controversial with the Afghan government. In 2007 and 2008 there were complaints that aid was being allocated according to security needs, with a primary objective of establishing security and winning the support of local populations for the central government. International civilian organizations became increasingly concerned about perceptions among local recipients who did not separate military and civilian delivery of aid. They were concerned that their civilian staff and Afghan civilians would be attacked by insurgents who might mistake receiving humanitarian assistance for assumed loyalty with pro-government forces (Jackson and Haysom 2013: 12).

Ranging in quality and accomplishments, PRTs established in 2003 Iraq were civilian-led interagency teams designed to provide a US "civilian face" to Iraqis "by bolstering moderates, promoting reconciliation, fostering economic development and building provincial capacity" (PRT Fact Sheet 2009). Each participating ISAF or coalition partner developed their version of PRTs with different emphasis on security, development, and civil–military integration. Their overarching goal was to promote stability and development while focusing on governance, national unity, rule of law, economic and political development at the provincial level. US PRTs included staff from the Department of State, USAID, Department of Defense civil affairs, US Department of Justice, US Department of Agriculture, US Army Corps of Engineers, subject matter experts and local Iraqis. At their peak they provided a primary regional connection between the US, coalition partners and provincial and local governments in every Iraq province. Seven additional e-PRTs were embedded with military brigades after the surge in 2007, and were phased out by 2010. Their priority was to serve as a platform that enabled the UN, National Democratic Institute, International Republican Institute, NGOs to support elections, serve the needs of women as well as other US government and international actors (PRT Fact Sheet 2009).

UNAMI viewed the formation of PRTs in Iraq with caution stating that while they shared similarities with humanitarian organizations their goals and objectives might be very different. They maintained that PRTs carried out a range of counterinsurgency strategies shaped by political and security goals heavily influenced by the US military mission and were viewed as the primary interface between MNF-I coalition partners and provincial and local governments in all of Iraq's 18 governorates. UN staff was encouraged to maintain a degree of awareness about PRT programs to avoid duplication, inconsistency, and minimize competition (Shearer 2008: 3). PRTs were disbanded in late 2011 and turned their projects over to the Government of Iraq. UNAMI, previously cautious about interacting with PRTs, agreed to continue support of specific projects (PRT Fact Sheet 2009).

Changing the civil–military conversation

The replication of little understood Vietnam types of civil–military models in both Afghanistan and Iraq make it clear that more research and discussion is needed to understand the impact of combining humanitarian and aid projects on civilian and military participants and local populations (see Stout 2013). Rather than examining underlying reasons for failures and identifying successes, the US tendency has been to eliminate an entire model if a strategy or program is not achieving results within a limited timeframe. The effect of pervasive corruption at all levels throughout US involvement in Vietnam constantly undermined efforts to achieve political, military, and social legitimacy in the eyes of the Vietnamese but was often underrated or ignored when evaluating civil–military projects (USAID 1975b: Vol. I, 144/237). This left those who followed to make decisions about past successes or failures of earlier strategies and programs based on partial information. Counterinsurgency programs designed to win the hearts and minds of local populations seem to be particularly susceptible to this process of partial knowledge and evaluation.

Recent and emerging conflicts have proved that each situation has its own set of complex issues that do not respond well to conventional civilian and military remedies (Johnson 2014). This was borne out in discussions of a hearts and minds strategy for Afghanistan and Iraq, which were often treated as proscriptive doctrine with little to support that assumption. Evaluating the history of counterinsurgency from Malaya to Iraq, Paul Dixon noted that contrasting conventional wars to "hearts and minds" counterinsurgency models conceals a range of very different interpretations. The levels of consent sought from the local population can run the scope between those who believe a population's loyalty should be won through coercion, fear and deference to those who seek a greater degree of popular consent, enthusiastic support and trust from the people (Dixon 2009: 363). The combination of North Vietnamese Army and Viet Cong insurgents created a specific type of civilian and military situation that did not exist in Malaya. While they shared similarities on the surface the situation in Malaya proved to be very different from political and military problems in Vietnam where civil–military public health and education programs designed to win the hearts and minds of the Vietnamese people proved to be "a goal which proved far more elusive than had ever been anticipated" (USAID 1975b: Vol. I, 21/278).

Experiences in Afghanistan and Iraq have challenged the US and international community to develop new ways for civilian organizations and military forces to work alongside each other in dangerous environments. Expectations by military forces, especially the US, of what civilian organizations could and should accomplish were unrealistic due to significantly lower numbers of civilian staff and smaller budgets. Both interventions also highlighted differences between US civil–military communities and their international counterparts. Internal US interagency relationships and those between the US military and international civilian organizations tended to be very different. International civilian humanitarian and aid organizations had long been engaged in different conversations about civil–military roles that did not always synch with US military objectives. They were much more concerned about establishing perimeters of humanitarian operational space than advocating for a "unity of effort." While this had long been an underlying concern within the US humanitarian and aid community (see Chapter 1) it was a strongly articulated worry among members of the international community. The "hearts and minds" counterinsurgency approach designed to help those populations considered to be most beneficial to the goals of the interveners was especially problematic for the international civilian community who perceived this strategy as violating a fundamental premise of aiding those who needed it in both Afghanistan and Iraq. Military forces also tended to "over" respond to situations by planning a mission before they had defined the problem (Cuny 1989). Others felt that international aid workers' assumptions that they could operate as neutral entities relying on the protection of the populations they helped was a misguided and naive strategy that could leave themselves, their local staff, and the communities they served more vulnerable to targeted violence.

There is cautious acceptance by members of the international humanitarian and aid community that continuing civil–military dialogue is necessary to negotiate roles and "space" acceptable to all. There have been calls for the US Army to have the serious objective debate they never had after Vietnam to better understand their experiences both good and bad that will allow them to create a force that has been "thought through and pre-pared for a broad range of possibilities" to adapt and respond effectively to changing conditions on the ground (Johnson 2014). The reluctance of civilian humanitarian and aid organizations to develop a coherent policy among themselves to guide their interaction with military forces contrib-uted toward making aid delivery more difficult and dangerous than it need have been. International civilian organizations and military forces in both Afghanistan and Iraq often did not include the host government or local organizations early in the decision-making process for programs and pro-jects that they were expected to continue.

OCHA drafted new guidelines for civil–military coordination in 2013. These stated that militaries could contribute to humanitarian action by rap-idly mobilizing and deploying unique assets and expertise when responding to specific requests and that coordination between humanitarian and military forces could range from cooperation to coexistence (OCHA would manage these activities through UN-CMCoord). They made the distinction between military actions that support political policies and humanitarian assistance that is provided according to need without taking political positions or sides in disputes stating that effective and consistent humanitarian civil–military coordination is a shared responsibility crucial to safeguarding humanitarian principles and operating space. Finally, humanitarian staff must be aware of problems that could occur from working with military forces and take steps to ensure that their neutrality, impartiality, operational independence and "civilian character" of their humanitarian assistance was not compromised (OCHA 2013b).

Delivering civilian and military assistance, skills and expertise to fast moving situations is necessary to achieve success in future interventions. Large scale international crisis interventions are evolving to keep pace with the different levels and types of civilian and military assets and inter-action required for different levels of violence. How military forces and civilian organizations interpret and negotiate their roles within a coherent civil–military approach will directly impact the success or failure of their efforts. In this era of budget cuts and geopolitics it is time for US military forces and civilian organizations to develop nuanced sophisticated strate-gies that address subtle but important issues. The following chapters in Part II explore key issues in negotiating future civil–military roles, protect-ing civilians, and understanding violence in a world where social media interacts with real time geopolitical violence on the ground. Despite per-ceived failures if the US and international civilian and military commu-nities chose to remember and heed the lessons from Vietnam and recent experiences in Afghanistan and Iraq, future civil–military efforts will be based on solid and more effective strategies.

Part II
Reframing the issues

3   Negotiating space

Finding common ground for negotiating civil–military roles is becoming increasingly important in a world where "unpredictability is the name of the game" (Guterres 2013) and communication technologies shape events in real time. Responding to multi-layered natural and manmade disasters known as complex crises and emergencies requires a level of response that is beyond the capacity of any single agency or organization. Characterized by violence, food insecurity, epidemics, conflicts, population displacement, and damage to social and economic systems they often occur in regions, countries and societies where there is a near or complete breakdown of authority, environmental degradation, terrorism, and destabilizing crime networks (International Federation of the Red Cross and Red Crescent Societies 2012). The introduction of "applied" information and communication technologies (ICTs) especially social media and web-based platforms during humanitarian crises and conflicts has compressed timelines and reshaped the ground rules for international civil–military relationships and interventions.

As the international civil–military community struggles to meet these new challenges it is clear that reacting to threats with solely military force does not deter others from emerging. While civil–military operations will most likely continue to downsize revisiting the lessons about civil–military models that evolved from Vietnam to the large scale operations in Afghanistan and Iraq offers insights into how to negotiate more effective civil–military roles and relationships in the future (see Chapters 1 and 2). The international community is beginning to realize that responding to these future challenges may require more, not less civil–military dialogue. Urged by their peers to accept the fact that "militaries will rarely have a purely humanitarian role" even when responding to natural disasters civilian humanitarian actors are nevertheless in a position to influence this dialogue by promoting the basic tenets of IHL, the international humanitarian law framework that seeks to limit the effects of conflict on civilians while seeking common civil–military ground (Haysom 2013: 3).

This chapter begins a discussion about how to reframe civil–military roles to respond to future challenges that continues in the remaining chapters. Recent civil–military discussions about experiences in Afghanistan and Iraq have highlighted significant differences in large scale civilian and military organizational capacity: numbers of personnel; capability: logistics support, the ability to travel in insecure areas, and financial resources as military budgets have overwhelmed civilian funding (Hartwell *forthcoming*). Acknowledging fundamental differences in organizational cultures and missions this chapter discusses how identifying these and other issues emerging from the long wars can help build constructive, positive relationships and strategies that move beyond these differences to achieve shared goals.

Understanding differences

Learning to understand differences between civil–military organizations begins by acknowledging that civilians and militaries bring different but not incompatible skills and strengths to planning and implementing projects in the field. Participation and observation during the military to civilian transition process in Iraq drew a clear picture of how many of these distinctions had been blurred during the large scale "whole of government" approaches favored by the US in Afghanistan and Iraq. These interventions normalized large scale "comprehensive" or "stabilization" strategies that tended to incorporate humanitarian and development activities into the military's "toolbox" of options as they pursued alternative strategies to achieve military goals (Metcalfe *et al.* 2012: 5).

Civilian strengths include the ability to initiate and implement sustainable development, humanitarian, and human rights projects; facilitate and support community-led reform and peace building; and lead teams that address political, economic, social, and health issues. Military strengths include a type of professional rapport with formal and informal armed groups; logistical capability to provide safe transport and safe work environments for civilian-led humanitarian and development projects; the ability to develop and implement security strategies for host country elections, public events, forums and to establish "neutral" safe spaces where civilians can interact to reform government institutions (Hartwell *forthcoming*).

Most civilian organizations who work in crisis areas can be loosely grouped into two categories, interagency government organizations that operate within a government's jurisdiction and a wide range of organizations that operate independently outside government control. Within these categories many organizations differ in how they set their agendas and priorities. US government affiliated civilian organizations include the Department of State, the US Agency for International Development (USAID, under the Department of State), and the related 'interagency' umbrella that includes the Department of Labor, Treasury, Justice, Homeland Security, Agriculture, Energy, Commerce, Transportation, and the Central Intelligence Agency—each with their own internal dynamics and cultures. For example, diplomats such as the US State Department Foreign Service Officers are trained to observe and report. While actively engaged with the host country's top leadership they prefer to consider a range of outcomes for longer periods of time before reacting. Now under the Department of State, USAID has had a different history (see Chapter 1) with staff who actively initiate and implement projects on economic growth, agriculture, trade, global health, democracy, conflict prevention, and humanitarian assistance. They understand the fundamentals of sustainable development but have lacked direct control over development projects initiated by the US military (Hartwell *forthcoming*).

Militaries working in the same regions tend to be large national forces with civil military operations that are assigned "missions" of foreign humanitarian assistance, population and resource control, national assistance operations, military civil action, emergency services, civil administration, and domestic support operations. Many of these include regional military alliances acting under unified commands such as NATO and United Nations peacekeeping forces (DPKO). All militaries operate with a clearly defined hierarchal system that excels in logistics and tasking and in performing a defined set of goals with anticipated concrete outcomes. As fighting forces they are ready to act at any moment.

Organizations outside the US interagency government framework operate on an international, regional, national, and local basis with a range of focus and goals. They are often referred to as intergovernmental or international government agencies, international non-government organizations, NGOs and civil society organizations (CSOs). NGOs and CSOs are both national and international. For example, the Department for International Development is a UK government agency working on a range of development-related issues throughout the world. The United Nations and the World Bank are independent international institutions guided by their country members with limited intervention peacekeeping, financial, judicial and humanitarian powers. Relief and humanitarian organizations such as CARE, Mercy Corps, Oxfam, Médecins Sans Frontières, and the International Committee of the Red Cross are part of a non-profit global community who work alongside government and international organizations in unstable environments. They are also known as NGOs, which operate on international, regional and national levels. CSOs tend to be local, community-driven non-profit organizations underwritten by local, national and international funds.

The United Nations was one of the first international organizations to promote a coherent aid, development, and security approach during multi-layered peacekeeping operations in 1990s Bosnia and Kosovo (van der Lijn cited in van de Kuijt 2012: 24). During this time civil–military cooperation operations and interaction increased as did a greater number of independent NGOs working in conflict areas (Van de Kuijt 2012: 21). Previously tasked with solely military missions and objectives, military forces became more directly involved in delivering and providing humanitarian aid as complex crises and emergencies increased. A comprehensive approach was adopted by the EU in its European Security Strategy of 2003 and NATO adopted an integrated comprehensive approach in its 2006 Riga Summit Declaration, in which they declared that "cooperation and coordination between organisations, individual states, agencies and NGOs, the private sector and the host government" along with contributions from "all major actors" was required to ensure security, stability, and development, which could not be achieved by solely military means (Van de Kuijt 2012: 21). Canada promoted a comprehensive whole of government 3D approach (diplomacy, defence, development) in Kandahar, Afghanistan during 2005. Building upon the 3D approach crisis management strategies and operations began to include political, civilian, military, government and non-government actors to evolve into "comprehensive" and "coherent" (Metcalfe *et al.* 2012: 5) civil–military interventions.

During this time civil military operations included missions focused on foreign humanitarian assistance, population and resource control, national assistance operations, military civil action, emergency services, civil administration, and domestic support operations. Regional military alliances acting under unified commands such as NATO and United Nations peacekeeping forces (DPKO) increased crossover into what had traditionally been civilian "turf." Governments and aid agencies who agreed to abide by basic principles of "humanity, impartiality, and independence" regardless of political affiliation had previously provided emergency humanitarian assistance in these situations. While international development funding had often been implicitly linked to political goals, such as helping to support specific political transformations, it was widely understood that donors should refrain from linking aid to a primary pursuit of their own political agenda (Oxfam International 2011: 7–8).

In a 2013 conversation at the US Embassy in The Hague, Netherlands, two Dutch NGO Cordaid's staff members viewed the principle of coherence positively but noted that the type and degree of cohesion varied between strategic and operational levels. Each had worked with the Royal Netherlands military (members of ISAF-International Security Assistance Force) in Uruzgan province, southern Afghanistan (2006–10) and observed that the greatest degree of cohesion was achieved while carrying out missions on the ground (Interviews 2013: The Hague, Netherlands). They shared recent papers that described key "dilemmas" or challenges. These included fundamental differences in timelines, culture, and capacities between diplomacy, defense, and development organizations, and the need for additional resources, time, effort, and funds as cohesion increased (known as a classic dilemma in cooperation). The long term sustainability of comprehensive approaches that depended upon short term political will for their support and execution was also problematic (van der Lijn 2011: 6). Five factors that determined the success or failure of civil–military projects were found to be the costs and benefits of participation; the alignment of institutional goals, mandates, and objectives; sharing organizational views and values; the specific context of the conflict; and individual "chemistry" between civilian and military personnel. The more these issues were acknowledged and "dealt with" the better the chances for success (van der Lijn 2011: 6).

A number of international humanitarian agencies who did not welcome sharing what they viewed as their traditional roles with military forces resisted these types of operations. In *Whose Aid is it Anyway* (2011) Oxfam International questioned the military and security goals of certain donors whose "skewed aid policies and practices" threatened to undermine a "decade of government donors' international commitments to effective, needs-focussed international aid" (Oxfam International 2011: 1). A lack of a cohesive policy among civilian humanitarian and aid organizations working with military forces also contributed to confusion over civil–military roles.

Military forces have become more experienced working with civilians in government and non-governmental organizations in their area of operations and are actively consulting with national and international organizations. There has been growing acknowledgement among international civilian aid and development organizations that greater clarity is needed on "key aspects" of civil–military relationships, especially the "principle of last resort" (humanitarian guidelines for requesting military assistance in relief operations) and information sharing protocols (Metcalfe *et al.* 2012: 29). Past experience has shown that some level of civil–military dialogue, preferably in the earliest phases has a "greater chance of preserving humanitarian space and influencing military conduct." Despite differences, varying degrees of cooperation, coherence, and co-existence have been successfully demonstrated across the civil–military spectrum. Even when civil–military relationships have been extremely tense efforts made by both sides have been effective in developing clear structures and mechanisms for coordination (Haysom 2013: 4).

Building civil–military trust

If civil–military organizations are to successfully negotiate future operational space it will be necessary to build a deeper level of understanding, communication and respect that goes beyond tolerating differences to accepting a range of analyses and approaches to complex crises and emergencies. Building the level of trust necessary to support the depth of organizational flexibility, thought, and measured action necessary to make careful decisions in chaotic circumstances will require international civilian and military organizations to address the following critical issues.

Acknowledging the key role leadership, personality, experience play in successful outcomes

In both casual conversations and interviews civilian humanitarian and aid workers and members of military forces have consistently emphasized the critical role quality leadership and the "right" personalities play in achieving goals and completing projects (Interviews 2013). There is nearly universal agreement that civil–military goals can be implemented in the most extreme circumstances when apparently incompatible leaders share the same focus and determination to succeed. Conversely, simple projects can be sabotaged by poor leadership and lack of interest. Leadership training, an integral part of military education, has been primarily focused on internal relationships or external relationships from a military perspective. Development of civilian organizational leadership both within and outside government has followed a similar internal focus. While many civilian and military leaders have successfully overcome their differences to achieve positive results, future challenges will require both civilian and military organizations to cultivate flexible, open minded leadership models to successfully negotiate their roles in unpredictable environments.

Experience is also an important factor in developing effective civil–military relationships. Leaders who listen to and learn from others, especially subordinates with prior experience in similar circumstances, will understand how to manage expectations, identify limitations, and communicate in tense environments. While it is difficult to change an individual's personality civilian and military organizations can teach their leaders how to diffuse tensions caused by personal differences and evaluate their potential impact on highly sensitive projects when making assignments. This level and quality of leadership, experience, and flexibility combined with the personal characteristics of civilian and military leaders will become especially important to effectively respond to and de-escalate crises when ICTs are compressing timelines for decisive action.

Understanding threats from civilian and military perspectives

Civilian and military organizations have often misunderstood each other's reaction in threatening situations. Unarmed civilian humanitarian and aid workers conducting operations in dangerous environments quickly learn to identify indicators of eminent violence and develop survival strategies to avoid it. On the other hand, military forces intentionally move toward and engage with violence to stop it. Some of the most interesting conversations I had with military personnel in Iraq were about civilian and military views of personal safety. Movements of civil–military teams within war zones in Iraq and Afghanistan often highlighted differences between acceptable levels of risk, especially among US forces and civilians without prior military background.

Military forces who are trained to identify, react, and engage with violent events as they occur tend to not recognize more subtle indicators of violence. They engage in direct combat and react with deadly force when they find themselves in dangerous situations. They are organized and equipped for travel in dangerous areas. As a result, military forces would often decide to undertake movements on the ground if they gauged threats to personal safety as high but manageable. Military forces view personal threats collectively and are trained to depend and fall back on the cohesiveness of their unit and team to protect each other under fire. They understand that death is a potential consequence of their profession and each member is assured that he or she will receive public acknowledgement of their ultimate sacrifice and "community" support for their loved ones (Hartwell *forthcoming*).

Civilians experienced in maneuvering inside dangerous areas will use different criteria to evaluate risks. Unarmed civilians working in war zones and unstable areas understand the inherent dangers but tend to take the view that intentionally putting yourself in harm's way is inviting unnecessary trouble. Civilians who do not have military backgrounds, especially members of diplomatic, humanitarian, and development organizations, do not carry weapons and are trained to mitigate threats by enacting individual strategies and taking actions that minimize personal risk. Unlike their military counterparts, civilians are not guaranteed a communalized public ritual or extended support if they lose their lives on the job. Understanding that there may be a possibility of death in dangerous areas, many civilians will view the violence they encounter as directly related to their circumstances. While there are increasing institutional pressures to serve in a hardship assignment their experience is similar to the types of violence experienced by other unarmed civilian researchers and practitioners (Hartwell *forthcoming*).

Prior to Iraq and Afghanistan, many American civilians and military forces had been unaware of the numerous unarmed civilian practitioners and researchers who were working in dangerous areas to provide emergency humanitarian assistance, develop institutional, social, and infrastructure projects and conduct fieldwork interviews without military escorts. This mostly international group developed extraordinarily sensitive barometers for "reading" violence and coping mechanisms that reasonably assure their personal safety. The United Nations High Commissioner for Refugees (UNHCR), one of the international humanitarian groups experienced in operating in unstable environments, have well honed personal protection strategies in place. Their *Handbook for Emergencies* (2007) warns personnel not to travel alone or after dark if it can be avoided or carry large amounts of money; to always tell someone your destination when leaving and the time you expect to return; to keep vehicle doors locked and windows rolled up when travelling, and to strategically park vehicles for a fast exit. Workers were instructed to refrain from taking photos of or around military personnel or installations and to have cash, documents, and an emergency bag packed and ready to go at all times. Above all they were advised to be polite and courteous to local officials, police, and military, as rude behavior might result in negative consequences for them and other members of their staff (United Nations High Commissioner for Refugees 2007).

International and US researchers and academics from a range of disciplines have long utilized ethical codes and personal protection strategies when collecting ethnographic interviews in destabilized environments. Similar precautions are taken by members of human rights organizations who document testimonies of war crimes and abuses in the field. A landmark book, *Fieldwork Under Fire: Contemporary Studies of Violence and Survival* (Nordstrom and

Robben 1995) documented anthropologists' personal stories and strategies for conducting fieldwork in areas affected by political violence. US anthropologist, Christopher Kovats-Bernat, who conducted long term fieldwork among violent gangs in Haiti listed strategies for understanding and adapting to threats. These include: listening to local advice on what is happening and where; organizing a safety support system that consists of calling in and/or showing up at an appointed time and place; developing specific strategies for very dangerous areas—no visible notebooks, recorders or note taking of any kind (this may include speaking on mobile phones); and most importantly understanding when you should terminate research and evacuate for personal safety (Kovats-Bernat 2002). Civilian and military interactions with the process of violence is explored further in Chapter 5.

Choosing visible and invisible civil–military interactions

Choosing when to meet publicly or visibly or discreetly or invisibly will become increasingly critical to protecting civilian workers, military forces, and local populations in the social media age. Over the past decade of long wars invisible meetings between international and local NGOs, other civilian organizations, and military forces have been held off-site or in discretionary local settings on an *ad hoc* basis. These off-the-grid civil–military interactions have taken place when visibility would have jeopardized a project's success and the safety of all participants. Vulnerabilities experienced by members of both civilian and military organizations underlie discussions about the visibility of their interactions. While many international humanitarian and aid workers would prefer to keep their distance from political and military actors, their perceptions of neutrality and the ability of local populations to protect them have receded as they become increasingly targeted by terrorist organizations. Military forces are also confronting the limitations of their force as they learn more about ways in which their presence can provoke rather than deter attacks (see Chapter 5).

To avoid harming local populations and themselves, civilian and military organizations may find it increasingly necessary to communicate and coordinate their movements earlier while carefully choosing visible or invisible interactions both inside and outside their areas of operation. Understanding the importance of sharing information and enforcing the principle of "do no harm" will go a long way in addressing their vulnerabilities. Civilians who carry out successful projects and programs with local populations can be attacked when military forces and government intelligence agencies intentionally or unintentionally use them as covers for intelligence gathering and targeting operations. Armed forces are vulnerable in situations where civilian organizations working alongside militaries do not share information they have received about impending attacks. Visible civil–military negotiations might include open cooperation and coordination between individuals and organizations. Invisible negotiations may appear on the surface as if no civil–military interaction is taking place while discreetly communicating and coordinating their movements. As technological acceleration of crises becomes the norm, careful decisions will need to be made on when the best results can be achieved away from public scrutiny in circumstances that do not unnecessarily endanger civilians or military forces.

Impact of information and communication technologies

The introduction of "applied" ICTs, social media, and web-based platforms during humanitarian crises and conflicts have begun reshaping ground rules for international civil–military relationships and interventions. The impact of instant communication about events in one part of the world on events thousands of miles away demonstrates how the developed and developing, urban and rural, stable and in crisis world are now connected in real time. Refugees and internally displaced persons in rural areas of the poorest countries could now access a mobile phone, use an internet café (set up with hardware purchased by refugee entrepreneurs or donations by international organizations), and watch satellite television. Urban refugees and internally displaced persons can also communicate with each other, international humanitarian organizations, and security forces through short message service (SMS) text messages, social media platforms, and phone hotlines (Aleinikoff 2011: 4).

These new technologies have already reconfigured global dynamics for the international humanitarian community. The increased participation of external, invited and unsolicited volunteer and technical communities (V&TCs) first in Haiti, then in Libya, has resulted in rising demand for rapid international responses from both military and civilian organizations to emergencies. The Haiti 7.0 earthquake in January 2010 was the first natural disaster to dramatically highlight the importance of ICTs as an interface between the international humanitarian community, online V&TCs and the Haitians they were trying to assist (FMR 2011). Early on it was clear that Haitian survivors were far more sophisticated in their ability to utilize communication and information technologies than were the international humanitarian organizations trying to help them (Wall 2011: 6). Survivors trapped under debris sent text messages for help while random individuals and CrisisMappers linked to shared networks OpenStreetMap and Sahana translated and mapped Haitian requests for assistance (Turner 2011). Disaster Relief 2.0 (2011) reported that the international humanitarian system was not ready to cope with the "two new information fire hoses" one coming from the community coping with the disaster and the other from a "mobilized swarm of global volunteers" in adapting to the "new reality where collective action can lead to collective intelligence" (Harvard Humanitarian Initiative 2011: 9).

The learning curve continued during the fighting between government and rebel groups seeking to overthrow Libya's leader Colonel Muammar el-Qaddafi in 2011. Learning from the Haitian experience, the United Nations Office for the Coordination of Humanitarian Affairs (UN OCHA), Geneva, activated a self-organized Standby Volunteer Task Force (SBTF) in March 2011 to create a Libya crisis map that tracked events as they evolved. The CrisisMappers' SBTF mapped social media, news reports, and official eye-witness reports from within Libya and along the borders. Coordinated by OCHA's Information Services Section the task force collected and mapped who what where (3W) information while the United Nations Operational Satellite Applications Programme (UNOSAT) hosted the Common Operational Datasets (Verity 2011; OCHA 2011).

Two versions of the map were launched. One contained sensitive information available only to supervised users and the other was a public version that obscured personal identities and descriptions of original reports while providing information already online (OCHA 2011; Verity 2011). An OCHA data specialist was assigned to examine incoming data for patterns and trends to determine how this information could be integrated into traditional coordination activities. The map site and data were made available to other organizations to assist with their operational planning. OCHA, UNOSAT, NetHope and the V&TCs—CrisisMappers, Crisis Commons, Open Street Map, and Google Crisis Response Team collaborated to document and map violent incidents in real time (Verity 2011). Assisted by volunteers from OCHA Columbia who had been recruited from the UN's internal online volunteer service and the SBTF who stayed beyond the first month's "deployment" the mapping project continued until June 2011 (OCHA 2011).

Unanticipated challenges rose from external virtually connected volunteers who preferred working directly with populations on the ground rather than helping to provide background information to agencies working in the field. Skyping, emailing, sharing and assessing risk management strategies continued non-stop. With 24 hour global connectivity the UN liaison was over whelmed by incoming data and difficult questions that needed fast answers. It became clear that organizational structures needed to be in place at the beginning of each project. The UN also learned to be more flexible in integrating use of non-UN standard software such as Skype and Google Docs. The Geneva-based OCHA information management team found Skype to be effective in collaborating with self-organized teams of volunteers. They organized Skype chats to integrate OCHA field staff with other international locations and sought the input of local staff to develop technological tools and software (Verity 2011).

Facing a future of reduced budgets and smaller forces, the US military has also reassessed their use of information and communication technologies. The Department of Defense declared that global counter-terrorism efforts would "become more widely distributed" and "characterized by a mix of direct action and security force assistance" that included "humanitarian, disaster relief and other operations" under the Counter Terrorism and Irregular Warfare section in their 2012 publication *Sustaining US Global Leadership* (Department of Defense 2012: 4–6). They offered communications support to civilian agencies leading relief operations as well as developing joint doctrine on military options that would prevent and respond to mass atrocities, providing assistance by air and sea including surveillance, medical evacuation, care, and non-combat evacuations for American citizens abroad, and to be on call for emergencies (Department of Defense 2012: 4–6).

Focused on more covert operations the US Special Operations Command (SOCOM) overseeing SEALs and elite Special Force teams has been stockpiling electronic media for air drops in remote locations. These include AM/FM broadcast transmitters and antennae, miniature loudspeakers; entertainment devices, "game device technologies" (Shachtman 2012); miniature power sources, hardware able to withstand harsh environments, internet broadcast reception and transmission devices, applications for cell phones that include text messaging, video, television streaming, and renewable power sources and rechargers that are lightweight, easy to track and identify (Shachtman 2012; SOCOM 2012: 13–14). These devices are designed to replace traditional methods of face-to-face conversations and printed leaflets previously used by the US to disseminate information (Shachtman 2012).

SOCOM has also created new user-enabled social media radio applications, described as a cross between talk radio and Twitter, a military version of friends-enabled Shoutcast in which users produce their own long-form radio shows by "dialing in with a free phone number" (Shachtman 2012). Designed to be downloaded onto PCs, Macs, and smart phones these applications would allow those on the ground to share their "thoughts," experiences, and interests in sports, music, news, and culture with each other and the US government (Shachtman 2012). An untraceable cellular device would function as an *ad hoc* network broadcast tower that supported and concealed the identity of friends and colleagues who posted messages, answered surveys, called in to a DJ/monitor, and listened to local, regional, and international broadcasts (SOCOM 2012: 13–14; Shachtman 2012).

Apart from secret operations, the military could share newer devices and technologies with civilian humanitarian and aid organizations working alongside military forces, which would offer a greater opportunity for crisis affected local populations to communicate imminent or potential danger. They would also provide secure platforms for seeking and organizing external international volunteers from the crowd sourcing, multiple sources, and V&TCs to relay messages, correlate data, distribute information, map crises and give a voice to "anonymous" and deliberately silenced populations. Disadvantages of their use are the lack of ambiguity or nuance in information flows that are easily manipulated and time consuming to vet and analyze. Open systems can be monitored by enemies who hack into networks to spread disinformation, and make sources more difficult to protect. Crowd sourcing external volunteers can cause coordination problems and provoke internal conflicts.

As real and virtual worlds rapidly converge there is an urgent need to continue developing more sophisticated civil–military protocols for validating and sharing information. Understanding the difference between factual information and analytical knowledge necessary to translate raw data into meaningful context has become critical. While reports in real time will enable faster and more precise rescue, aid, and protection operations, information itself will continue to have minimal value without context, perspective and thoughtful analysis or the political will necessary to respond.

Looking ahead

Civilian and military organizations can no longer afford to ignore each other in dangerous environments. Reframing familiar issues that include developing information management and governance strategies; developing protocols for ethical use of social media; publicly acknowledging and supporting different styles of democracies, and learning to ask the "right" development questions will be increasingly important to respond effectively to future crises. Equally essential will be to accept the limitations of organizational capacity and capabilities dictated by numbers of personnel, budgets, geopolitics, and instability. Civilian and military organizations need to better identify how and where their different skills, experiences, strengths, and vulnerabilities can be best utilized. In some circumstances civilian organizations will be more effective working directly with civilian populations while security forces will be better communicating with and influencing formal militaries and informal armed groups.

As fast moving challenges continue to disrupt the world order it is clear that a new set of global pressures is unfolding. The long term impact of information and communication technologies on future civil–military roles and relationships is unknown but growing. It is important that civilian and military organizations agree on how and when to share knowledge that saves local civilian lives without compromising their own operations or personnel. Developing information management and governance strategies that include a mutual civil–military protocol for sharing information on threats directed at civilian humanitarian and aid workers or members of the military would go a long way toward addressing mutual concerns. These management and governance strategies could include identifying potential sources to gather information through crowd sourcing, standing up in country platforms; negotiating information protocols including vetting information and protecting sources; governing the use of technology in the operational environment and monitoring the quality and sophistication of analysis. Utilizing a range of visual communication technologies from Skype to sophisticated video teleconferencing systems, supporting virtual interaction between like-minded groups through intergovernmental or transnational platforms for exchanging information and experiences between politicians, experts, bureaucrats, individuals, militaries, humanitarian and aid workers would broaden discussions.

There is a pressing need to develop civil–military protocols for the ethical use of social media that will monitor unfiltered information flows generated by communication and information technologies in violence-prone environments. Events in Haiti provided a glimpse of how information and communication technologies might impact future natural disasters, but Libya demonstrated the continuing dilemma of finding ways to protect individuals and their communities during its use. While these technologies are new the challenges they present have long dominated the "do no harm" and human security debates in international civil–military discussions. For most civilian organizations the use of new communication technologies in emergencies must avoid crossing the line from monitoring humanitarian consequences of conflict-related events to what could be perceived as military reconnaissance. Both civilian organizations and military forces could agree to a formal understanding that any information putting civilians or military forces in direct danger will be quickly revealed to the targeted party.

Joint civil–military criteria could also be established to vet and validate incoming information and data as it pertains specifically to current operations, choosing when to use visible or invisible military assistance to support humanitarian work and identifying how these technologies interact with violence on the ground. Protocols could be established by interviewing civilians and military forces with field experience. Ethical issues that rose during the Libya Crisis Map project were protecting the identity of Libyan sources on the ground, labeling the source of data as anonymously and vaguely as possible, and ensuring that none of the mapping activities inflicted harm on specific Libyan individuals or communities. OCHA solved this problem by running two separate websites. One site held all the original data accessible only to pre-approved users, and the other was a public site cleaned of identifiable information with data posted following a 24 hour delay (Verity 2011). Legal issues concerning privacy, liability, and intellectual property are emerging that challenge existing legal and policy frameworks, which will need to be adjusted, clarified and reconstructed to address the role of these evolving technologies (Shanley *et al.* 2013: 867).

Public acknowledgment by the international civil–military community that there is no one "right" version of democracy would help support development of credible institutions in newly democratizing countries. This stance would help to manage expectations through clear communication between interveners and the host country and encourage bottom-up peace efforts. This would also help to reinforce a new government's legitimacy, which depends on negotiated deals between citizens and their government to oversee credible security operations, support ethical host nation and international leadership, and promote accountability and transparency in the media. Legitimacy of interventions has been defined by the degree to which local populations accept and support missions and initiatives, the quality of formal and informal governance structures, and positive reinforcement by regional neighbors and the international community (USIP/PKSOI 2009: 3, 16–18).

Over the past decade military forces have experienced steep learning curves in trying to establish sustainable development projects in Iraq and Afghanistan. There is an opportunity for militaries to lay the groundwork for long term projects as they are often the first outsiders to access local populations in recently stabilized areas. Training by experienced development practitioners could assist forces in identifying problems and learning how to ask the "right" questions of local populations. Responses could then be communicated immediately to humanitarian, development, and health organizations with the expertise to address them. This approach would also give military civil engineers an opportunity to begin reforestation, anti-erosion, and simple sanitation projects designed to support long term environment and health-related infrastructure projects. Punching bore holes for wells they could leave confident that local populations will keep them operating as it is in their self-interest to maintain access to clean water (Graham 2013).

An unintended consequence of military development projects is that they can negatively impact ongoing conversations between civilian development agencies and host governments in countries where they are active. Former USAID official, Jim Graham, stated "from past observation and experience in Rwanda and other African countries the government stops development dialogue the minute the military moves in" (Graham 2013). Cordaid, the Dutch NGO working with Royal Netherlands military forces (members of ISAF-International Security Assistance Force) in Uruzgan province, southern Afghanistan, reinforced this view by recalling their difficulty in establishing civil–military integration and coordination with regional "host nation actors" whenever external military forces increased civil–military integration on a national level (van der Lijn 2011: 6). This is highly problematic as there is consistent evidence that the most sustainable projects result when external civilian organizations collaborate with a host country's NGOs and civil society organizations to ensure local investment in a project's success. As short term military operations increase, it will be important for both civilian and military organizations to define development roles that will work best for local populations.

While lessons learned from operations in Iraq and Afghanistan are important it is the shared experience in developing civilian–military relationships that will be increasingly valuable in meeting future challenges. Chapters 4 and 5 discuss how these relationships will help protect vulnerable civilians and develop more coordinated strategies to deter violence. Chapters 6 to 8 reinforce the importance of the key issues discussed in this chapter in reframing international civil–military conversations and responses to meet a range of new challenges. Marshall Adair, a Minister-Counselor in the US Department of State, emphasized the importance of civil–military relationships in his memoir, *Lessons from a Diplomatic Life* (2013). Recalling his POLAD (Political Advisor) role with SOCOM headquarters in Florida, 2003–2006, as embodying a "certain cultural challenge" in attempting to address the very different perspectives between soldiers and diplomats, he maintains that the skills of both "must be used together and in balance" as it is important that both professions continue to "ensure the health and effectiveness of that cooperation" as international threats increase (Adair 2013: 208). Accepting differences, focusing on common goals, developing creative problem-solving skills and inviting wider input into decision-making processes will facilitate more effective responses within politically driven technology-fueled environments. Complex crises and emergencies will require more sophisticated analysis, evaluation, and operational techniques. While there are underlying commonalities, accepting at the onset that there is no "right" answer for every situation will require that civilian–military relationships and operational frameworks be renegotiated on an ongoing basis to result in better outcomes for all.

4 Protecting civilians

An enduring frustration and tragedy in this era of instant communications is the difficulty in protecting civilians. Earlier expectations that violence would stop once a peace process began have been unrealistic. Instead civilians are often targeted by increased or continued score settling, assassinations, bombings, kidnappings, domestic violence and rape, which create unstable environments that support formation of new militias and gangs, transnational criminal activities, and finance terrorism networks. These chaotic environments look very different from the "old wars" fought between each country's army in cross-border conflicts that resulted in clear winners and losers. Instead large areas of the world have had to endure chronic insecurity where repeated cycles of violence and recurring violent civil wars have become a modern norm that cannot be ended with a clear cut victory or a negotiated settlement (Kaldor 2012; World Development Report 2011: 57).

UNHCR's High Commissioner Antonio Guterres has warned of unprecedented global humanitarian crises and challenges emerging in a world where international power relationships are unpredictable and the international community is unable to stop conflicts (Guterres 2013).

This chapter gives an overview of geopolitical challenges to civil–military interventions attempting to protect civilians. In this context civilian protection strategies are civil–military actions taken to provide basic survival needs and security to protect populations being targeted or displaced by intense violence. While displacement can result from natural disasters it is more often caused by ethno-sectarian violence directed at specific groups, equally affecting uninvolved bystanders caught in the middle. It builds upon issues raised in Chapter 3 to suggest that the international civil–military community could develop more effective protection strategies and interventions by integrating civilian perceptions of security into planning and missions. This could be accomplished by developing a more nuanced understanding of civilian identities in violent environments, engaging with non-state actors to protect civilians, and developing better understanding and support for the potentially important role that communication technologies could play in protecting civilians.

Geopolitical challenges to protecting civilians

International interest in developing specific policies to protect civilians increased in the late 20th century. The UNDP's 1994 Human Development Report defined human security as an international development concept that separated the security and wellbeing of individuals from the larger security of nations and encouraged governments to include civilian protection in their human security policies. In 2000 a Canadian-led International Commission on Intervention and State Sovereignty developed an interpretation that evolved into an international policy known as the responsibility to protect (R2P). This view advocated the responsibility and right of the international community to use force when necessary to protect vulnerable populations from genocide, ethnic cleansing, massive human rights violations, and the failure of their own governments to intervene on their behalf (Beebe and Kaldor 2010). The United Nations built upon this concept in a 2008 report on security sector reform that outlined a "people-centred approach to security" where they defined security as a situation in which all have "the right to live their lives and raise their children in dignity, free from hunger and the fear of violence, oppression or injustice" (Schroeder 2010; DCAF 2008: 4).

While there is often agreement on the need for some type of "hard power" to protect civilians there are major roadblocks to carrying out its timely implementation. The rise of globalization and communication technologies at the end of the "Cold War" has had the effect of reducing state autonomy and making older conventional military threats and their counter-defenses less relevant (MacFarlane and Khong 2010: viii; Thakur 2010: viii). These changes accompanied by increases in technologically advanced weapons and shifts in attitudes toward civilian casualties have made civil–military protection strategies increasingly difficult to implement in violent environments.

Predatory governments

One of the greatest challenges to protecting civilians is from predatory governments who violently attack their own civilians to consolidate economic and political power. While it would appear that protecting civilians would be in a government's best interest to strengthen internal security and project a positive public image, instead many mobilize powerful military groups, corrupt police, judges and prominent politicians to carry out assassinations, bombings, ethnic cleansing, forced internal population displacement, genocidal and sexual violence against their own citizens (Wulf 2006: 29; Thakur 2010: viii). These governments also target civilians by working with national military forces to form alliances of convenience with non-state armed groups that enables them to gain resources and territory. The extent of the killing may be obscured by government officials who launch internal debates and external public relations campaigns while calculating which actions they can take without triggering external interventions. Civil–military interventions may be limited or not take place as these governments stop short of genocide. In other cases potential interveners may view targeted groups as shared enemies and choose to do nothing about the abuses (De Waal and Conley-Zilkic 2006).

Hugo Slim has defined a government's potential for violence against its civilians as limited, limitless, and inevitable. "Limited war" is described as a rare form of restraint and protection that respects human life while simultaneously acknowledging that war is inherently chaotic and dangerous and can unintentionally harm civilians. While civilians dislike the outcome they tend to be more understanding and willing to forgive accidents in these circumstances (Slim 2008: 22–3). "Limitless war" is on the opposite end of the spectrum where a violent winner take all philosophy views force as a positive "cleansing and creative" agent of change. Proponents of this strategy actively seek to purge society of undesirable individuals and groups while simultaneously consolidating their power. All civilians in the designated "undesirable" group are seen as justifiable targets (Slim 2008: 25–8). "Inevitable war" is shaped by a type of pessimistic resignation that sees war as a regrettable but unstoppable force resulting in unavoidable civilian casualties. This view emphasizes a perception that the human condition is under a "hellish curse" that will never respond to the ideology of restraint (Slim 2008: 29–30). It views attempts to protect innocent civilians as essentially futile. Only "pacifism" stands distinctly apart as an ideology that rejects all types of war while actively seeking its prevention and abolition. However, once a war is underway pacifists often pragmatically seek to limit its destructive effect on civilians (Slim 2008: 31).

UN Secretary General Ban Ki-moon emphasized the need to pressure nations to comply with international rights and humanitarian law to effectively protect women, girls, boys, and men during a 2011 Security Council speech. He urged interveners to implement targeted sanctions and intense scrutiny against the offending countries and to consistently engage with non-state armed groups to improve their compliance with international laws. He suggested that the international community could help strengthen national security and legal institutions by supporting development of a well trained proactive peacekeeping force that had the ability to work with national military, police, justice, and corrections systems to reinforce human rights principles. Advocating for increased cooperation between aid organizations and groups in conflict to improve humanitarian access to distressed populations, Ban Ki-moon called for perpetrators to be held accountable and given serious punishments for their offences (Ki-moon 2011).

Political will to intervene

How members of the global community react to a predatory government's violence against its own civilians is driven by each country's political will to intervene that is often shaped by national and regional political, economic, social, non-secular and other strategic interests. Many internal conflicts have complex positive and negative implications for foreign governments.

Ethno-sectarian violence playing out inside the borders of one country may offer opportunities for other nations to shape outcomes in their favor by staying silent or engaging in proxy warfare. These strategies can take the form of open or clandestine support by supplying weapons to designated groups, sending in fighters, providing economic assistance, or expressing public condemnation while resisting external interventions. Incentives for directly supporting a country's conflict may include interest in eliminating common enemies, changing regional power paradigms, promoting political and religious ideologies, or reprioritizing national interests that are indifferent to the outcome. This behavior reinforces the observation that "stopping mass violence has rarely been a high priority for those foreign governments that have the capacity to intervene" (De Waal and Conley-Zilkic 2006).

Arguing that the first increasingly multi-polar world to emerge in one hundred years is reshaping the global power paradigm, Stephen Hopgood points out that earlier international endorsement of responsibility to protect policies lulled many into a false reassurance that targeted and mass killings of civilians would be stopped by future civil–military interventions. Many of these assumptions were based on Western European concepts of secular-based human rights law and philosophy that are less relevant to the types of cultural and non-secular beliefs reflected in recent power shifts. This dilemma is exacerbated by the waning influence of Europe and an increasingly ambivalent United States struggling to balance its own political and economic interests. As driving forces behind the dominant human rights and international justice model, the US is facing new challenges by emerging (Brazil, India, South Africa) and rejuvenated (Russia, China) regional powers who have dire human rights records. Similar to the US and Western European countries these nations have their own agendas and interests that challenge prevailing interpretations of transitional justice and humanitarian interventions, which have often been perceived by them as a type of victor's justice. Emerging global powers may choose similar types of civil–military interventions to protect civilians but they are likely to do so for different reasons, setting the stage for a different type of human rights and justice model (Hopgood 2014).

Civilians protecting themselves

The idea of civilians evaluating the effectiveness of strategies meant to protect them is a relatively new concept for the international community. Until recently most protection policies have been developed from the top down without seeking the opinions of locals who often found themselves caught between foreign civilians and militaries in the day and militias, terrorists, and criminal gangs at night. Distinguishing civilians who are innocent bystanders from ones who pose grave danger to security forces and civilian interveners is far more difficult in unstable environments than widely understood. In conventional wars fought across national borders civilians and combatants were easy to identify but most contemporary conflicts now take place inside a country without international legal policies in place to distinguish or codify the difference (McDonald 2004: 2–3).

Much political and military skepticism toward civilians has been less focused on who carries weapons than on their perceived sympathies, encouragement, and support for individuals and groups and how these are reflected in their political, economic, and social activities (Slim 2008: 209–10). However, this viewpoint is too limiting as it does not take into account how violent processes evolve and impact affected populations. Once a conflict begins the parameters of civilian identity become frozen, rigidly drawn and defined by his or her enemies and reduced to a simple brutal premise—you are simply *not* whomever or whatever your enemy *is*. If armed groups attack civilians, fighting back or cooperating with their government's enemy may be necessary to survive. In the midst of intense violence civilians can be coerced or motivated by ideology, faith and survival into overtly or silently aiding enemies of security forces assigned to protect them. They will be forced into multiple revolving roles where victims become perpetrators and perpetrators turn into victims. Others will enable or protect groups by remaining silent about their activities or by providing support such as food and shelter to avoid attack. These often include vulnerable women and children who are observed or perceived to be aiding armed groups (Slim 2008: 184,188–9, 209–10; Gorur 2013: 4; Hartwell 2005: Chapter X).

Coping strategies may put civilians in a difficult position with civil–military interveners as they can lose their right to protection under international law if it can be shown they have directly participated in hostilities. Many civilians are willing to risk that chance as they develop their own systems for evaluating the effectiveness of external civil–military interventions in dire circumstances. To implement effective protection strategies external civil–military interveners need to become more knowledgable about how civilian identities are shaped during ongoing violence and the range of self-protection strategies civilians find most effective when countering different types of threats. (See Chapter 5 for discussion on coping with violence.)

Civilian perceptions of security

A range of self-protection strategies employed by individuals and communities when attempting to counter, mitigate, deter or avoid a threat was described by African participants in a survey conducted by the Stimson Center's Civilians in Conflict Project in the Democratic Republic of Congo. These include utilizing local defense groups and community patrols to deter or confront perpetrators, seeking violent retaliation against specific offenders as a form of rough justice, and seeking a companion or group to accompany them while traveling to, from, and during work. Holding regular meetings with local officials to discuss security priorities, plan strategies, share timely security information and develop a warning system with each other and other communities was important. Participants described a range of useful tactics that included publicly denouncing and testifying against offenders, utilizing civil society organizations to report infractions to political authorities and act as their advocates, staging protests by refusing to open businesses, and attempting to resolve the conflict and reconcile adversaries. Some circumstances forced survivors to evade and hide from attackers or to flee and resettle after an attack while in other instances submitting to and cooperating with armed groups would avoid attacks. When no other options appeared to be available some resorted to praying for protection (Gorur 2013: 4). These measures were used to counter ongoing sexual violence, robbery, pillage, targeted killing, land conflicts, arbitrary arrest, illegal detention, and extortion carried out by levying illegal and exorbitant taxes, fines and tolls (Giffen 2013: 17).

A UN commissioned study Protecting Civilians in the Context of UN Peacekeeping (2009) outlined recommendations for improving the effectiveness of UN peacekeeping missions and their partners in protecting civilians. These included linking members of the Security Council to the field, improving mission-wide strategy and crisis planning, improving the role of uniformed personnel and focusing on political follow-up in achieving mission aims. It was recommended that the Secretary General and the Department of Peacekeeping Operations ensure that responsibilities for protecting civilians were clearly detailed in the Secretary General's directives and compacts between the Secretary General and the special representatives of the Secretary General who speak for the Secretary General at important meetings on human rights issues. Member states were urged to draw upon their own experiences to develop more effective peacekeeping missions to protect civilians (Holt *et al.* 2009: 14).

Communication technologies' potential for self-protection

An emerging option for civilian self-protection is the use of information and communication technologies that includes social media, mobile phone networks, and web-based applications to communicate potential threats and escalating violence. Wireless and internet-based communications technologies are not only changing international civil–military relationships but have been proved to be effective in giving local populations the means to directly communicate their needs during natural disasters and conflicts (see Chapters 3, 7, 8). The 2010 Haitian earthquake demonstrated that disaster survivors were far more sophisticated in their ability to utilize communication and information technologies than were the international humanitarian organizations there to assist them. Haitians trapped under debris sent text messages on mobile phones pleading for help while concerned individuals around the world linked to OpenStreetMap, Sahana, and CrisisMappers to share networks that translated and mapped requests for assistance (Wall 2011: 6; Turner 2011). While Haiti provided the first model for ways in which information and communication technologies might affect future civil–military interventions in natural disasters Libya put their use for protecting individuals and communities squarely in the limelight. Increased participation of external, invited and unsolicited volunteer and technical communities during that crisis resulted in local and international demand for more rapid responses from both military and civilian organizations. Future crises are likely to combine elements of Haiti, Libya and an array of additional destabilizing factors into "complex human emergencies" (Turner 2011) with the potential to engulf entire regions.

A 2011 Disaster 2.0 report observed that individuals using "cloud," "crowd," and SMS-based technologies had participated at an unprecedented number in responding to disasters. When working together traditional relief organizations, volunteers, and affected communities could "provide, aggregate and analyze information" that sped up and improved humanitarian relief (Turner 2011). While incorporating these communication technologies into wide ranging protection strategies is in its earliest stages they have significant potential for local populations to inform others of their need for protection and deter threats in real time.

Developing effective civil–military strategies and interventions

One lesson drawn from the wars in Afghanistan, Iraq and elsewhere is that conflicts evolve and change over time. State and non-state actors may trade roles or new global and local actors may emerge to reshape circumstances bearing little resemblance to the original violence. This scenario challenges many earlier civil–military protection policies designed on the assumption that conflicts would evolve in a reasonably linear progression from violence to peace. "Spoilers" who sabotage peace processes for personal gain are not new (Stedman 1997) but perpetrators of modern political and criminal violence form convenient transnational alliances to infiltrate local conflicts, social protests, and gang violence to facilitate criminal syndicates and international terrorism. This violence flows over borders to endanger a range of countries with different incomes, national identities, religions, and ideologies (World Development Report 2011: 66).

While understanding how civil–military roles were implemented in Iraq and Afghanistan will continue to be an important lesson, it is the process of developing those civil–military relationships (see Chapter 3) that will be increasingly valuable in developing effective civilian protection strategies in the future. Civilian protection will require more sophisticated analysis, evaluation, and operational techniques by international civil–military interventions. Deciding when and how to use "hard" power interventions or "soft" power public diplomacy with social media will require an adjusted mindset and greater awareness of the potential of unintended negative consequences. Ongoing communication between civilian and military interveners will need to combine nuanced observations with sophisticated analysis in combinations of high and low tech strategies that successfully addresses potentially volatile complex crises and emergencies (see Chapters 6 to 8). Success will require a degree of focus, observation, understanding, coordination, and flexibility that is difficult for the international community to achieve.

Meeting these challenges will require that future civil–military interveners consider and incorporate the following in preparing, planning, and launching their mission.

Integrate civilian self-protection strategies into civil–military planning

Civil–military interveners will need to incorporate a summary of real and potential threats, vulnerability assessments of different groups, deeper analysis and more nuanced evaluation to increase effectiveness of protection strategies. External interveners could use this information to clearly define missions and goals that focus on terminating or minimizing identified threats, decrease the vulnerability of the targeted groups, and track and analyze unintended short and medium term negative consequences directly caused by the intervention (Giffen 2013: 6). For example, while wealthy merchants in towns may be the primary target of armed groups their presence may increase looting and violence throughout the region. Displaced populations camped close to villages may be more vulnerable to attacks if they have tense relationships with the villagers who will not directly warn them when they are too far away to hear the whistles of their early warning system. In these situations civil–military interveners and peacekeepers could encourage host communities and displaced populations to organize security meetings to develop mutual protection strategies and provide them with updated equipment to improve their warning system (Giffen 2013: 11).

Engage with religious leaders and faith-based NGOs to support protection strategies

Unlike secular nations that officially separate religion and politics religious institutions and affiliated NGOs are the major influence on shaping political power in non-secular societies. Many of these religious leaders are concerned about the wellbeing of their constituencies but their potential to make a positive contribution toward protecting their followers and others has been regarded with ambivalence by the US and developed world powers. As the authority of major global secular institutions formed after World War II (including the United Nations, World Bank, and International Monetary Fund) begins to wane, the input of contemporary emerging faith-based global players is increasingly important as these movements gain influence in shaping new interpretations and models for human rights and justice. Religious influence differs from conventional political power as it gives moral weight to related claims and arguments in a way than cannot be achieved by secular-based politicians and politics. While this shift requires compromise many religions have more nuanced interpretations of these issues than is commonly assumed by many secularists. Ongoing questioning and resistance to protection policies based on Western European traditions implies that a new more multi-cultural faith-based interpretation of human rights is emerging and with followers being informed by social media and information technologies (Hopgood 2014). If the international civil–military community engages religious leaders and their followers by using private and public diplomatic strategies their input can provide positive opportunities to lend moral weight to efforts that strongly influence their followers' cooperation in protecting their fellow civilians.

Engage with non-state actors to protect civilians

Non-state armed groups operating on their own or in areas alongside military forces present other dilemmas for implementing effective protection strategies. Forming alliances of convenience with other armed rebels, terrorists, and criminals they often develop external ties with international groups linked to predatory governments that create additional threats to civilians. While many groups operate under some type of code clarifying violence against civilians their interpretation of which civilians are exempt from attack varies according to circumstances. In an open reply to a request by the UN Assistance Mission in Afghanistan (UNAMA) in 2012 the Taliban declared that civilians such as "white-bearded people, women, children and common people" who were living an ordinary life and not involved in the fighting should not be attacked or killed but anyone working with security companies or escorting foreign supply convoys were legitimate targets (Casey-Maslen 2013).

There has been concern in humanitarian circles that too few civilian organizations attempt to protect civilians in conflict areas dominated by non-state actors. Proposed rules of engagement for negotiating humanitarian principles with armed groups include understanding why a group may not be willing to comply with accepted international definitions of civilians, conversing with the widest possible range of armed groups without endangering civilians, initiating contact by building upon pre-existing or mutual relationships, and being clear that a groups' participation in negotiations will not guarantee that their legal status will change under international law. Civilian organizations should identify and monitor "windows of opportunity" to initiate negotiations, identify and build upon positive incentives to reinforce a group's cooperation, and portray an impartial role while monitoring and acknowledging when civilians are targeted. Groups should be informed about their obligations under customary international humanitarian and human rights law while agreements should be documented in written form whenever possible and groups encouraged to develop, adopt, and communicate a code of conduct based on these principles (Casey-Maslen 2013). While civilian organizations may be more effective in protecting civilians in some circumstances, there are others when military to military negotiations between formal and non-state forces may prove to be more effective in protecting and creating safe spaces for civilians and humanitarian activities, as armed groups share a mutual professional understanding and rapport (see Chapters 3 and 6) that can be advantageous in dangerous environments (Hartwell *forthcoming*).

Develop and share new communication technologies

There is little doubt that the country controlling the electromagnetic spectrum that enables wireless communications, navigation, logistics, and virtually guided attacks will gain future military superiority. As these technologies become less expensive and more accessible to any state or non-state individual and group who desire them attempts to control this spectrum are evolving into a new type of arms race (Koerner 2014). While sophisticated weapons and cyber warfare rely on this technology there is a case to be made for the range of options they offer vulnerable populations to protect themselves by responding to stop threats spread by disinformation and inflammatory rhetoric. Encrypted communication technologies that do not need a dependable source of electricity or internet connection can also be used to report movement of armed groups to locals and the global community. The Zello walkie talkie application, which works on a range of mobile phones, can be encrypted and used on any wifi or data plan as has been demonstrated during 2014 activity in the Ukraine and Venezuela. Other apps such as tumblr send instant photographic evidence of attacks as they occur. Many of these technologies have been adapted by local populations in recent years to communicate threats to members of international protection, humanitarian, and technical communities.

Understanding the difference between factual information and analytical knowledge is critical. Advantages and disadvantages of using these technologies are discussed in Chapters 3, 5, and 6. Information flows are factual descriptions of time, place, and actors while knowledge is developing an understanding and analysis of the links between local and international populations and ways in which these technologies interact with processes of violence and peace, societal crisis, natural disasters, and destabilizing influences. Civilian protection principles that respond to the use of these new technologies need to be developed and agreed to by decision-makers who oversee intervening civilian organizations and military forces. Civil–military planners can evaluate these data flows by comparing their information on civilian self-protection strategies and conflict identities to pinpoint the most effective strategies that protect vulnerable civilians in real time.

Reprioritizing efforts

Developing effective strategies to protect civilians requires a reordering of international civil–military priorities during interventions. For many peace support operations and military missions this will require developing flexible guidelines that anticipate complex and dynamic changes on the ground. The use or non-use of force to protect civilians needs to be "clearly spelled out" to find an "appropriate and effective compromise" between implementing massive fire-power and doing nothing (Herbert Wulf in Glasius and Kaldor 2006: 28). While hard power can stop violence and humanitarian organizations can help local populations survive only political solutions can end and prevent the majority of conflicts that will ensure sustainable protection of targeted civilians (Ki-moon 2011). Chapter 5 continues this discussion by depicting violence as a process, why it is important for interveners to understand that, and how it exposes and exploits vulnerabilities among both civilians and militaries.

Unfortunately the best laid plans for protection of civilians will be useless rhetoric if there is a lack of political will to stop the violence. As Syria and other conflicts prove, international civil–military readiness and instant information cannot compete with geopolitical agendas that support government forces and armed groups who violently target civilians. When military peacekeepers and civilian humanitarian and aid organizations are finally allowed access to these targeted populations they may find themselves attempting to protect cynical violence-hardened civilians who have learned through difficult lessons that they can only rely on themselves for protection. In these cases it will be critical that civil–military interveners respect and incorporate civilian perceptions of effective security measures into their protection strategies.

5 Coping with violence

Learning to cope with and survive violence requires a different mindset and set of skills than learning about a country's culture and language. It is a familiar but complex process that easily weaves itself into a country's institutions and social fabric. As deliberate targeting of civilians becomes a primary tactic in the "new wars" successfully averting future threats will require the international civil–military community to better understand how state and non-state actors, public and private spheres, and internal and external actions interact to directly affect local violence. Violent environments generally share two sets of characteristics, one with endemic violence lying just below the surface of daily life that appears to spontaneously erupt when the "right" combination of motives, circumstances, and timing converge, and one where urban and rural fighting between state and non-state forces takes the form of ongoing skirmishes and battles in the midst of civilians (Willman and Makisaka 2010: 46).

This chapter builds upon issues discussed in Chapters 3 and 4 to analyze violence both traditionally in what Johan Galtung described as critical issues—the use of violence and the ways in which its use is legitimized (Galtung 1990: 291) and by utilizing recent frameworks developed by the World Health Organization (WHO) and others that depict violence as a health issue with characteristics of an infectious disease. Military forces have much to learn from unarmed international humanitarian and aid workers who survive dangerous environments by correctly interpreting signals of impending violence and refining a mix of "hard" and "soft" security skills (Humanitarian Practice Network 2010: 112). Motives and the impact of social media and communication technologies are examined. In a virtually connected world where global disinformation can instigate local attacks, developing a commonly understood set of indicators and methods to understand and identify the evolution of violence will be essential to avoid unintended consequences in future crises.

Understanding violence

There is a wide range of motivations for violence. Cycles of violence generated by conflicts often evolve into sustained political, economic, and socially motivated processes. Political violence can occur when violent struggles between armed groups and their government are resolved by a military victory or negotiated settlement. Perpetuating violence may also be in the interest of politicians seeking to preserve and consolidate their power by skillfully manipulating ethno-sectarian violence under the guise of safeguarding a race or religion. Economically motivated violence uses homicides, kidnappings, and extortion to raise funds for individuals, illegal groups, and governments in cases of extreme corruption. Members of armed groups who have less to gain in peace than they did in war may continue to violently consolidate their control over areas by engaging in criminal enterprises. Social violence can be instigated by youths and gangs motivated by ethnic divisions, revenge, gendered targeting, and domestic violence (Harborne and Sage 2010: 5; Hartwell 2006).

There is a strong correlation between violence, natural resources, conflict, and disasters. Violence is attractive for its effectiveness in economic "resource wars" where endless fighting provides an opportunity to access illegal diamonds or additional sources of otherwise unattainable wealth (Keen 2012: 195–235). Approximately 40 percent of internal violent conflicts in the past 60 years have been linked to illegal acquisition of natural resources. The probability of a peace process relapsing into violence increases when exploitation of resources causes further environmental damage or when benefits are unequally distributed (UN DG 2013: 140–1).

Benjamin Ginsberg argues that violence is the driving force behind almost all political transformations and acquisition of power. He cites Thomas Hobbes who regarded violence as a rational way to access territory, safety, and glory, and Mao Zedong who believed political power came from the barrel of a gun and that politics was a continuation of violence by other means (Ginsberg 2013: B6). He makes the case that those willing to use violence to achieve their goals have been more effective in dominating or overturning election results, negating parliamentary laws, ignoring public opinion, and acting as a major agent of social and political change. They have set the stage for state and non-state actors to trade roles or create opportunities for new global and local actors to emerge as violent groups who are often only defeated by more powerful enemies who use greater force against them (Ginsberg 2013: B7–8). While there is much evidence to reinforce this premise it does into take into account the reality that unleashed violence can acquire a life of its own beyond the control of an individual or group. While violence may prove beneficial for perpetrators and bystanders in the short term it rarely has long term political benefits.

Psychologically motivated violence can be used to achieve utilitarian aims or simply be enacted for its own satisfaction. There are psychological benefits such as achieving or preserving status through intimidation and fear. Shame has also been suggested to be a strong motivator in perpetuating cycles of indiscriminate violence that have little in common with achieving dominance or "winning." David Keen maintains that shame-driven violence offers an immediate opportunity to act against feelings of powerlessness and injustice. Recruiters for terrorist, insurgent and other non-state armed groups can easily manipulate shame in potential fighters by justifying and rationalizing acts of violence against innocent victims. In this context external criticism or threats can motivate perpetrators to continue rather than cease their violence. Keen maintains that perpetrators can feel remarkably little shame or guilt viewing their acts as justified or "righteous" while victims can feel both. This situation sets up a revolving victim–perpetrator cycle where victims violently retaliate to regain their sense of control and respect (Keen 2012: 195–7). This and other types of revenge, generally understood to be driven by rage, are familiar motivations for violence. Separate from a victim–perpetrator cycle it can also be intertwined with a process of forgiveness and sense of fairness connected to how justice processes and legal systems are perceived to be functioning during crises and transitions. Desire for revenge can have a positive function when it is acknowledged but not acted upon by former enemies who make the decision to cooperate with each other to successfully rebuild political, economic, social and legal institutions (Hartwell 2005: Chapter 5).

Traditional analytical frameworks

Traditional frameworks for analyzing violence have been generally organized into three categories: structural, direct, and cultural. Structural violence incorporated into legal and social systems and inside institutions designed to favor the needs of some groups at the expense of others can lay a foundation for active or direct violence. Direct violence also known as secondary violence is the act of carrying out physical or psychological harm and is the type most commonly recognized. Cultural or symbolic violence utilizes icons and symbols found in religion, art, and ideology to justify, reinforce, and legitimize structural violence. Symbolic and culturally specific language, empirical and formal scientific processes (logic, mathematics), visible stars, crosses, crescents, flags, posters, military parades, and inflammatory speeches and anthems, are used to make direct and structural violence "feel" right or to at least "not wrong" (Galtung 1990: 291).

While violence is interpreted within a wide range of frameworks it shares universally recognizable characteristics in that it appears as an ephemeral process connected to a series of random observations and events. In reality these are clear signals of escalation or deterrence within different timelines. Galtung compared the process of direct violence to the theory of how earthquakes occur. He described the appearance of violence as a sudden "event" caused by structural violence similar to earthquakes, which occur from tectonic plates moving underground due to the potential volatility of underlying fault lines that function in the same way as cultural violence (Galtung 1990: 294). Analysis of this process has been constrained by professional "silos," which have prevented deeper understanding of the ways in which different types of violence reinforce and connect to each other. While relationships between different types of violence have been acknowledged they have rarely been analyzed through a larger lens. For example, a rise in domestic violence has consistently been observed as group violence subsides, but they have been treated as two separate issues resulting in a lack of cohesive policies that would have addressed their underlying connection (Willman and Makisaka 2010: 53).

Reframing violence as a health and development issue

Recent international efforts to eliminate myopic policies have reframed violence as a global health and development issue. These have been accompanied by an understanding of the very different benefits gained from a "negative" peace, which is the sole absence of violence versus a "positive" peace where the absence of violence is accompanied by good governance, accountability, and equal access to resources (Denney 2013: 5). Seeking to reinterpret a wide range of interpersonal violence, suicidal behavior, armed conflicts, psychological threats and intimidation, WHO published a 2002 report that included all acts of violence within a global health framework. Psychological harm, deprivation and underdevelopment that compromised the wellbeing of individuals, families and communities were given the same consideration as deaths and injuries (WHO 2002: 4). Direct violence was re-categorized to include self-inflicted violence such as suicide, substance misuse and other types of self-harm; interpersonal violence carried out by an individual or small group against an individual, stranger, or family member (domestic violence); and collective violence such as organized crime, wars, and terrorism enacted by organized groups to advance a political, financial, or social agenda (Willman and Makisaka 2010: 52). Acknowledging that international moral codes made distinguishing acceptable from unacceptable harmful behaviors a complex and subjective issue, WHO proposed a basic framework that integrates analyses to develop a broader range of prevention strategies (WHO 2002: 4).

The NGO Cure Violence has taken this analytical framework one step further by depicting violence as a contagious disease that exhibits three key characteristics "clustering, spread, and transmission" of an epidemic similar to HIV, cholera, and Ebola. They maintain that violence is a changeable behavior, which the health sector experienced in dealing with similar issues such as addictions and harmful behaviors (smoking, eating, etc.), and is in a position to best understand, diagnose and treat. This approach attempts to reduce or stop violence by identifying and treating those most at risk for this behavior in the same way that health professionals identify and treat those at risk for specific diseases. Mapping and tracking violence in the same way as the spread of disease allows the utilization of real time interventionist approaches by carefully selected community members and trusted insiders. Trained to anticipate where violence may occur they are educated in techniques to mediate ongoing conflicts, prevent retaliation, and keep older conflicts from re-igniting. Framing violence as a health risk emphasizes the value of prevention as a more effective and fiscally sound approach than attempts to treat the more devastating results (Cure Violence 2014).

The United Nations Development Programme portrays violence as an international development issue linked to root causes of conflict and violence that include heightened civilian vulnerability, climate change, unemployment, financial instability, poor resource management due to poverty, poor governance, and social insecurity. It is identified as the primary cause of instability linked to the lack of personal security and ability to live a peaceful life, and can create national development deficits such as inequality, marginalization of women, children, and youth and illegal use of natural and state resources. Protection of women and girls has been especially important in the face of mounting evidence that a lack of equality and gendered focused violence prevents the international community from stopping conflicts and sustaining long term development growth. Efforts are being made to include increased access to security and trust in the judicial process, provide fair and equitable opportunities for economic and political participation, and facilitate access to social services. Resources are also allocated to address the needs of youth whose exclusion could motivate their participation in criminal activities and other violence (UN DG 2013: 139–40). This view prioritizes a civilian development response over security forces that are only called upon if they are needed to stop violence. Designing effective prevention strategies has been challenging as violence is "intensely context specific" and what proves to be a successful deterrent in one context may provoke violence in another (Willman and Makisaka 2010: 46). The intertwined and unpredictable nature of violence reinforces the case for developing a multi-dimensional civil–military approach to its prevention.

Surviving violent environments

Avoiding attacks in violent environments will increasingly require a mix of hard and soft security skills. Hard security skills, similar to those learned by military, police, and intelligence services range from technical expertise in operating equipment, alarm systems, body armor, blast film and walls to understanding military tactics, using weapons, conducting investigations, and analyzing threats and risks. Soft security skills include the ability to analyze and understand the context of violence within a range of cultural, social, and political environments. This includes developing and maintaining community information networks to receive alerts for potential threats as well as the ability to mentor, train, communicate, plan, and build relationships and budget skills within multi-cultural teams (Humanitarian Practice Network 2010: 112).

As discussed in Chapters 3 and 4 civilians and military forces differ in how they respond to violent threats. The type of violence that civilian teams encounter will dictate how they will balance and use different security skills. Militaries emphasize the training of hard skills but there are fewer efforts to educate their forces on understanding violence as a process or to offer opportunities to develop soft skills that can avoid unintended consequences for others and themselves. In combat environments civilian humanitarian teams include a greater number of members with hard skills while soft networking skills may be in greater demand in very tense areas where tribal and local militias are not fully engaged in open warfare. High crime regions may require an equal mix of both (Humanitarian Practice Network 2010: 112–13).

Unlike most military forces civilian humanitarian and aid organizations categorize the types and level of threats by international, national, local staff and gender. While national and international staff are regarded equally in terms of duty to protect, in reality international organizations have limited options on actions they are able and prepared to take to protect these workers during crises. Many international agencies still provide the majority of organizational security, training, safe housing, transportation and communication for their international staff. They often evacuate their international staff when security risks are very high based on the assumption that national staff face fewer risks. Growing numbers of attacks against national staff have challenged this view as ties to the local area may increase rather than diminish their vulnerability to targeted violence. As national staff has become 85–95 percent of international aid agencies' global workforce in the past 15–20 years this is now a significant problem. Many have been with their organization for five or more years and provide valuable institutional memory and field experience. They may be members of the local population, from a different region in the same country or foreign third-country national workers who have been legally hired and contracted as local staff. Many local staff will eventually be reassigned as international staff in other countries (Humanitarian Practice Network 2010: 120–1).

Advantages from a security perspective for using international staff include a broader understanding of organizational values, culture, policies, and procedures, better understanding of donor and project management and reporting requirements, more objectivity in assessing environmental risks, and less susceptibility to being pressured by local actors. They are more likely to be perceived as impartial by groups in conflict and can easily liaise with international staff in other organizations. Disadvantages may be unfamiliarity with the background and history shaping perceptions and attitudes of local populations and less in-depth knowledge of local social, cultural, and political relationships. They may also lack insider access to local media or information networks and be more likely to engage in inappropriate or insensitive behaviors (Humanitarian Practice Network 2010: 114).

National staff have deeper knowledge of environments and history that shape local attitudes, which is an advantage for understanding regional violence. They speak local languages with access to local information networks, are in an excellent position to liaise with national staff in other organizations, and are accustomed to functioning in high risk environments (Humanitarian Practice Network 2010: 114). However they share the same vulnerabilities as the civilian population, which may put them at risk in deeply divided societies. Working for international organizations where they may handle assets, money, influence staff recruitment and award local contracts gives them status, visibility and wealth by local standards that increases their potential for being targeted in dangerous areas. Threats against them may include conservative social and religious groups who accuse them of associating with international organizations that promote values seen as disrespectful of traditional customs and norms and pressure by locals to provide jobs and benefits. They may be coerced into joining armed groups, risk being arrested by the government on suspicion of supporting an opposition group, forced by the government and other groups to provide intelligence and information about the organization and accused of aiding a foreign occupier through collaboration or spying (Humanitarian Practice Network 2010: 121). Accustomed to operating in this environment, it may not occur to them to raise the alarm about risks that they take for granted. They also may not be perceived as objective or more motivated by economic opportunity rather than the mission of the job—a possibility also shared by international staff (Humanitarian Practice Network 2010: 114).

Many civilian agencies now include gender stereotyping and gender-specific threats in their risk management strategies. Inflexible attitudes can contribute to internal team tensions and lack of focus, which could increase vulnerability in dangerous circumstances. Men may pose security threats in situations where they are likely to be perceived as aggressors and greater threats than women if they refuse to admit they lack necessary skills such as the ability to operate a four wheel drive sports utility vehicle in dangerous areas or if they adopt a "macho attitude" or a strong leadership demeanor that could increase tensions in volatile situations. While women have the potential to diffuse situations they are also more vulnerable to sexual assault and other gender-specific attacks (Humanitarian Practice Network 2010: 121–2).

Kidnapping threats to civilian humanitarian and aid workers

Kidnapping has grown into one of the greatest international threats against all categories of civilian humanitarian and aid workers. Calling kidnapping "the new normal" the 2013 Aid Worker Security Report warned that it had surpassed shootings, serious assaults, raids, bombings, and explosives as the most common significant attack against civilian staff working in violent environments (Hammer *et al.* 2013: 3, 5). Fourteen percent of the 372 kidnappings on record between 1997 and 2012 resulted in fatalities, with 80 of those occurring during abduction, captivity, attempted escapes, and rescues. Police or military forces rescued a small number of victims but the majority were released following unacknowledged private ransom payments by families, employers, and their home governments (Harmer *et al.* 2013: 4). The relatively low cost for carrying out kidnappings balanced against the potentially large monetary rewards and high profile political impact has made it attractive to both criminals and ideologically motivated armed groups. International staff are preferred targets as they are often easier to identify and guarantee greater international media and political attention. Kidnappers tend to hold international staff longer, for an average of 53 days compared to 12 days for national staff (based on 1997–2012 data), while demanding higher ransoms and concessions that require complex negotiations for their release (Harmer *et al.* 2013: 5).

Most kidnappers attack and ambush aid convoys moving on roads where they are vulnerable and easy to target. Opportunities for kidnapping and equally serious threats can occur when passing through official and unofficial checkpoints where interactions can be difficult and violence is quick to escalate. These can be staffed by unprofessional heavily armed individuals who may be tense, drunk, or under threat themselves. They may harass, intimidate and threaten workers or hold partisan or negative views toward the organization and staff attempting to pass through. How staff behave at these points is critical to insure their personal safety. It will also influence the future treatment of civilian workers from that agency as well as overall perceptions of the organization from that point onward (Humanitarian Practice Network 2010: 172).

The impact of kidnappings on individuals and organizations is devastating. The size, capacity and available resources of international organizations shape their responses. The United Nations has the capacity to activate a crisis management response system while continuing to operate in areas where the kidnappings took place but smaller NGOs may need to severely curtail or cease their operations to focus on the kidnapping (Harmer *et al.* 2013: 8). The greatest challenge occurs when individuals from different countries and organizations are kidnapped together and held by the same captors, making a collaborative approach that includes input from a range of home governments, families, and organizations harder to develop. The reactions of governments may vary widely from sensitive but strong support to taking opportunities to grandstand and meet ransom demands without taking into consideration the potential consequences for other programs and organizations operating in the same area. In almost all cases these kidnappings evolve into a prolonged crisis that requires managing a process of identifying and negotiating with the kidnappers, which may include multiple groups if victims are sold by their original captors to another group, diplomatic engagement with victims' families, their governments and the host country, orchestrating a release or rescue, and follow-up care for victims and their families. All this is likely to take place in the glare of public and social media, making protecting sensitive information a challenge and important priority (Harmer *et al.* 2013: 8).

Social media, communications, and violence

Interactions between social media platforms, communications technologies and unstable complex environments are emerging as one of the least understood factors, making these regions increasingly dangerous for civilian staffs and military forces. Social media can bring viewers closer to a conflict but it can also push them farther away. Without gatekeepers or filters these communication networks can spread false rumors, use coded inflammatory language and send unverified photos that can provoke immediate attacks on individuals or groups located thousands of miles away. Traditional photos and stories have the same potential for provoking reactions and staging false images but there is more time to vet and understand the back story behind the pictures and words. Constantly viewing a conflict through a Buzzfeed list or Instagram filter can make it seem less real or what cultural theorist Jean Baudrillard calls "hyperreal," an exaggerated version of reality that blends reality and simulation together to create a shared experience. Recorded protests posted online can feel more authentic than experiencing them first hand. Tweeted messages and Instagram photos lack the context of history, culture, politics and relationships that underlie a violent event and can be co-opted by those determined to instigate ethno-sectarian violence (O'Hagan 2014). Social media technologies can also be used by armed groups to intimidate and recruit and by predatory governments to shape internal and global political will or provoke targeted violence (Editors 2012: 4). While it is tempting to view public discourse as a "soft" power that establishes links and facilitates international dialogue, it is worth noting that the battle of values and ideas dominating public diplomacy during the late 20th century Cold War era led to more not less competition between hard powers (Melissen 2005: 5).

Recognizing the advantages of positive perceptions both international civilian organizations and military forces have developed strategies to control their public image with varying degrees of success. Many international humanitarian and aid organizations have taken advantage of different forms of media to raise awareness about a crisis, advocate for causes and increase an organization's visibility and profile with a view to raising funds. However increasing an organization's international visibility may decrease trust between them and the local communities they are attempting to help, especially when headquarters and staff working inside crises areas have conflicting goals. A poorly or inaccurately worded statement can put field staff in immediate danger (Humanitarian Practice Network 2010: 159). Military forces face similar messaging issues—especially the US Army (who developed a Media Relations Division in their Army Public Affairs Office) when using language such as cultivating "hearts and minds" or "protecting local civilians" when carrying out combat missions (see Chapters 2 and 4).

Revisiting the basics of good public relations by identifying clear goals and developing better understanding of the intended audience will be critical to conveying unambiguous messages that establish credibility (Morrison quoted in Waddington 2014). Despite social media's accelerated speed many basic strategies developed for vetting traditional news sources still hold. Good staff security practices published by the Humanitarian Practice Network (2010) recommend that an organization agree in advance as to how all members will officially respond to crises while taking care not to blame individuals or groups as the underlying dynamics of violence are not usually the sole fault of one group. All publicly shared information should be vetted by reliable sources or clearly state if it has not been to avoid spreading inaccurate and potentially inflammatory rumors (Humanitarian Practice Network 2010: 161). When sharing "off the record" background information staff should first confirm that journalists are professional, objective and agree to mutual ground rules including how the information source will be described. Many titles and descriptions of anonymous sources such as a "senior UN source" or "agencies operating in the conflict area" may be easy to identify. In a non-crisis situation it may be best to reach out to respected local, national and international media as a way to positively influence local communities and leaders (Humanitarian Practice Network 2010: 162, 159).

Establishing sophisticated communication equipment, networks, and training to manage communications during a crisis is essential for the safety of all civilian workers operating on the ground in dangerous environments. During violent events and crises such as a major security incident (improvised explosive device triggered by a mobile phone) or a natural disaster (tsunami, landslide) communications systems may be overloaded or deliberately shut down. Supplying teams with a range of communications equipment and training all personnel how to use it during violence and emergencies will help ensure their safety. Mobile phone networks, satellite phones, and satellite accessed internet, voice over internet phone technology including Skype, and GPS has made tracking movements of teams easier for organizational headquarters and security personnel but may also increase their vulnerability. The increasing technological savvy of non-state actors, criminal networks and others who may be able to hack, monitor or steal the equipment may require different strategies in highly insecure environments especially where radios and satellite phones may attract thieves or where GPS devices and digital cameras could increase suspicions.

Protocols for language use in distress and security calls during extreme emergencies should be established. The rule of thumb is to assume that if an exchange is not taking place face to face that it can be listened to. Learning to express oneself in a "moderate … non-partisan" manner and using code words to refer to offices, people, routes, vehicles, cargo, etc., and to share information on position or movements are key survival skills. Using metaphoric phrases to share sensitive information such as references to the weather for communicating political events or an attack works well but must be changed often to prevent the code from being cracked. If crime is a primary concern certain types of high value equipment will need to be used with discretion. Some technologies are also viewed with suspicion. Several NGOs working in Somalia found that GPS devices were increasing security risks when they were accused of using them to help military forces locate rebel group locations (Humanitarian Practice Network 2010: 141–2, 152, 158–9). In a world driven by social media networks and platforms monitoring drivers of internal violence and understanding ways in which they converge with external instigators will be essential for both civilian and military personnel to survive future interventions in complex crises (see Chapters 6 to 8).

Preventing and deterring violence

Preventing and deterring violence is a highly nuanced undertaking that requires focus, observation, and a clear understanding of the ways in which different types of violence are linked. Many early research and prevention activities were developed in isolation from each other, making many interventions too limited to be effective (WHO 2002: 35). Policies were often focused on controlling and preventing direct violence without directing resources or attention to violence rooted in specific socioeconomic and cultural conditions. This resulted in myopic approaches that ignored larger environmental and structural drivers of conflicts leaving the potential for repeating future cycles of violence (Willman and Makisaka 2010: 46, 152). The UN recommends basic policy goals that include attempts to eliminate all forms of violence against vulnerable women, children and groups, measure reduction and prevention of violent deaths per a benchmark of 100,000 population, and that focus on reducing external drivers such as human trafficking, arms, drugs, illegal sales of natural resources, money laundering mechanisms. They should also publicly support law enforcement and justice processes that are perceived by all groups to be legitimate, accessible, impartial, non-discriminating and responsive (Willman and Makisaka 2010: 4).

Developing common indicators of violence

To successfully address underlying causes of conflicts civil–military planners will need to agree on goals, data collection and assessment methodologies to measure levels of internal violence. Identifying motivations and sharing their analysis is a crucial first step toward developing protocols that avoid unintentional provocations especially when operating with limited information in rapidly changing environments. This requires planning civilian and military actions by using a commonly shared index of violence indicators and confirming current circumstances through direct civil–military communication whenever possible. Data can be collected from factual records that cite when, where, why and by whom violence is perpetrated. These could include police reports, hospital and rehabilitation center records, registries and other official sources to help identify types, trends, frequency and location of violence and identities of victims and perpetrators. Injury surveillance systems such as "Injury Surveillance Guidelines" developed by the World Health Organization in 2007 and household surveys that have been successful in identifying previously unreported incidents of sexual, domestic and other types of interpersonal violence are used widely. Official crime and violence observation groups established in many countries draw data from death certificates and other vital statistics records, hospital records, police crime statistics, court records and population surveys. A regularly updated and shared large scale database can be created by triangulating information from these sources to gauge the level of internal violence (Willman and Makisaka 2010: 4, 48–9).

If official resources are unavailable data can be collected from whatever information there is to assess the context, related problems, and local capacity to launch prompt and effective interventions. As violence and its effects rarely evolve in a linear process it is essential that information from formal and informal rapid assessments, small scale surveys in hospitals and other locations, questions about targeted victims and other simple baseline surveys be collected and carefully assessed. These smaller scale assessments are easier and more cost effective to conduct on a regular and as-needed basis while providing opportunities for practitioners to update their analysis and link issues relevant to other environments. Information from these assessments should be shared rather than having local leaders spend time and energy on providing the same information to each organization (Willman and Makisaka 2010: 48–9).

As any one assessment is incomplete it is critical to gather information from outside sources that monitor external relevant social and international media statements. These external data can be compared to an internal database that identifies possible violent reactions in areas where civilian and military interveners or local populations can be targeted. Future assessments of violence will be increasingly linked to independent disaster risk indexes that measure the effects of sustainable development, climate change, and conflict on poverty reduction, health, environment, governance, food security, gender equality, education, and water issues (UN DG 2013: 141).

When gathering data it is critical that civil–military analysts avoid substituting assumptions for missing information. Experience has shown that it is far better to identify and accept information gaps, acknowledging that some facts may never be known, than to plan civilian and military interventions based on assumptions that result in well intentioned actions having fatal consequences. Many humanitarian interventions during post-conflict transitions have given the greatest amounts of aid to those most in need assuming this would stop continued violence, but have instead found that these groups were often on the losing side of a conflict and assisting them was perceived by the victors as a political act that helped their enemy. For example, giving aid to Hutu refugees following the Rwandan genocide caused great resentment among Tutsi and moderate Hutus who had remained in Rwanda and viewed this international action as rewarding the perpetrators of the genocide (Willman and Makisaka 2010: 51).

Re-evaluating links between civilian kidnappings and violence

A growing challenge for negotiating future civil–military roles is to develop a mutual understanding of the ways in which the kidnapping of civilian organizational staff is linked to the type of violence occurring in shared areas of operation. Civilian worker kidnappings have been shown to rise sharply in regions where chronic violence has escalated into major combat, low-intensity conflicts, or asymmetrical warfare, with aid workers being treated as proxy targets. These types of kidnappings decline sharply once major tactical operations have ceased. Gaza and Sri Lanka led in numbers of kidnappings in 2009 that significantly declined once combat operations ceased. By 2012 Afghanistan, Pakistan, South Sudan, and Somalia consistently ranked among the most violent and dangerous areas for aid workers and operations with Syria joining them the same year (Harmer *et al.* 2013: 3). The correlation between open combat environments and higher numbers of kidnappings needs to be urgently discussed by civilian organizations and military forces. Raising awareness and negotiating strategies that will leave civilian staff of international organizations less vulnerable to targeted kidnappings that are the direct result from military operations will become increasingly important.

Conducting more nuanced assessments of security risks are essential for civil–military teams under military supervision and military forces operating in the same areas. Local populations observed to be working with the military face threats similar to the staff of international humanitarian and aid organizations. Preventing opportunistic kidnappings and violent incidents at checkpoints may require increased awareness by military forces of how their presence or actions may motivate kidnappers to attack international civilian workers. Military forces have tended to be less sensitive about using members of certain ethnic groups or diasporas for advice or translation in areas where their presence may provoke strong negative reactions. Adopting more focused strategies, using risk assessment tools compatible with ones used by international civilian organizations, and developing mutually acceptable civil–military protocols to identify the least provocative actions in a range of scenarios may help diffuse tensions and avoid violence. This would reinforce civilian protection policies that terminate or minimize threats, decrease the vulnerability of the targeted civilians, and track, analyze, and mitigate any harm that external interveners may cause to civilian humanitarian staff and local populations in the short and medium term (Giffen 2013: 6).

As civilian kidnappings escalate the Aid Worker Security Report (2013) has recommended that road travel security policies be re-evaluated and a collective security strategy developed between humanitarian and private security firms to seek solutions beyond adding more armed guards. Donors could be requested to fund innovative ways to keep workers safe while traveling in dangerous environments. They recommend additional communication, information sharing and lessons learned from analysis of the motives, context, and consequences of earlier kidnappings be reviewed with colleagues and other civilian organizations working in the same or similar environments. While there is a great deal of information on guiding a team response to kidnapping crises there has been far less focus and discussion on determining the range of endgames and strategic paths that may be necessary to release hostages (Harmer *et al.* 2013: 11).

Balancing short term mitigation with long term prevention

Balancing the needs of short term mitigation with longer term prevention strategies can be challenging during crises. Governments and civil–military interveners may be under intense pressure to produce visible, timely and measurable results, but experience has shown that sustainably reducing violence can take years and the participation of succeeding generations. Intense discussion within the international development community has focused on whether greater development gains can be achieved by emphasizing prevention of violence or conflicts that cause a type of "un-peace" or by concentrating efforts to achieve any type of peace at any cost (Willman and Makisaka 2010: 52). Even when political will to intervene is strong it is difficult to justify long term cost effectiveness for investing resources. While these distinctions appear to be minor each assumption forms a different set of international expectations, which in turn shapes international policies, interventions and how measures of effectiveness are defined and implemented by the international civil–military community (Denney 2013: 5). Staying focused on realistic expectations is essential as sustainable long term results usually require programs and projects that last a minimum of 2 or 3 years (Willman and Makisaka 2010: 47). When capacity and resources are limited interveners need to prioritize strategies that will most effectively address motivations for violence to achieve immediate security gains while simultaneously addressing underlying structural factors.

Successful violence reduction strategies have combined short term, quick win programs that incorporate targeted policing or urban upgrades with long term media campaigns to change destructive cultural norms. Policy-makers and civil–military groups have invested in rehabilitating violent perpetrators and re-integrating them into their communities. The UN Peacekeeping Mission in Port-au-Prince, Haiti, combined police patrols with sustainable community development projects in problem neighborhoods. This approach gave youth gangs responsible for the majority of violent crime in urban Haiti alternatives such as employment training and short term jobs that reduced their availability for recruitment by gangs (Willman and Makisaka 2010: 48).

Hitomo Kubo (2010) recommends a preventative rather than reactive strategy that would require a top-down, bottom-up cooperative approach between individuals, communities, and state forces that would set criteria for developing processes and institutions perceived as legitimate. Building upon Kubo's suggestions these approaches could incorporate perceptions of vulnerability and fear expressed by affected populations into a protection strategy. This would allow individuals and communities to identify and prioritize their security needs while providing insights into how and where threats occur. Military forces would be able to narrow their focus to develop an integrated macro–micro security policy in which hard force is applied to specific threats while underlying causes are continually monitored. This approach addresses a wide range of cross-border threats while taking into account the root causes of earlier and current conflicts that include weak institutions and fluctuating political, economic, and social conditions (Kubo 2010: 32–7).

Looking ahead civil–military interveners could negotiate roles that reflect maximalist and minimalist approaches that regard the absence of violence as a key factor in reducing poverty and creating safe spaces that facilitate political, economic, and social development. A maximalist or macro approach lists a range of "development disruptors" including violence, conflict, disasters, economic shocks, pandemic diseases, and the effects of climate change. The specific inhibiting effect of each disruptor on development and poverty reduction would be clearly identified, analyzed and incorporated into future policies (Denney 2013: 7–8). A minimalist approach would frame violence and conflict into politically feasible language such as "negative" peace that would prioritize the relationship between insecurity and poverty. This would require measuring perceptions and levels of violence, crime, injustice and the capability of services to address them. Measuring levels of corruption, accountability, etc., an essential part of a "positive" peace development strategy "for which weak, or at least contested, evidence exists" (Denney 2013: 8) would not be necessary unless included in a larger analysis of governance and institutions.

The international development community has advocated for sustainable management of natural resources to support vulnerable and conflict affected populations on a local level while spurring national economic growth and creation of jobs. A new development model designed to reduce all forms of risk with flexible "forward-looking" goals might include strategies to build internal resilience (UN DG 2013: 139–41). There has been a call to reframe future development agendas that take into account multi-dimensional risks and complex challenges to provide integrated development support. Attempting to eliminate corruption and bribes by holding public and private perpetrators accountable, enhancing state capacity, instituting more accountable transactions of national resources, and reducing socioeconomic inequality across groups and geographical regions could be included in short and long term policies. Achieving these goals would increase public trust, promote freedom from fear and encourage a sustainable peace that would guarantee a stable and safe society for urban and rural groups who endure disproportionately higher levels of disabilities, marginalization, and targeted violence (UN DG 2013: 141).

As new war scenarios integrating criminal and insurgent activities continue to evolve against a backdrop of urban and rural settings it is especially important that civil–military interveners develop a shared understanding of how violence evolves, how it interacts with communication technologies, and agree on protocols that prevent potential flashpoints and share life saving information. These approaches will require international civilian organizations and military forces to raise their interactions to more sophisticated levels than many have previously experienced.

Part III
Looking ahead

6 The "new war" challenge

The operating environment is changing fast. By 2015 the large scale civil–military experiences in Afghanistan and Iraq already felt like distant memories. Military and civilian leaders talk of volatility that is out of control, the "worst they've seen it in forty years" (Odierno in Sterman 2015; Guterres 2013). Pushed by rapidly evolving communication and other technologies the line between war and "not war" continues to blur (Rosa Brooks in Sterman 2015) underscoring Mary Kaldor's earlier warnings about the "new" wars with their perpetually chaotic environments and no clear winners, losers, or negotiated settlements. No longer fought within national boundaries wars would evolve into regions of chronic insecurity where cyclical violence and recurring civil wars would facilitate the formation of new militias, gangs, and financial networks linking transnational criminal activities with terrorist networks in mutually beneficial arrangements. Fast moving conflicts would become the new normal and with the advent of sophisticated social media and communication technology users capable of launching disruptions and violence at any time in any place throughout the world (Kaldor 2012; see Chapter 4). The recent humanitarian crisis caused by the violent mass displacement of Syrian civilians is just the tip of the iceberg. As personal dangers for civilians increase UNHCR's High Commissioner Antonio Guterres has warned about unparalleled humanitarian crises erupting on a global scale in a world where international power relationships are unpredictable and the international community is unable to stop conflicts (World Development Report 2011; Guterres 2013).

This chapter describes the importance of monitoring areas of potential trouble and why identifying and understanding potential causes of complex crises is crucial in an era when random events and technologies can combine with natural and other disasters to instigate unpredictable disruptions and violence at any time, any place throughout the world. This chapter sets the stage for Chapters 7 and 8 by describing current "new" war scenarios and anticipated challenges that will form future conflict environments and cascading crises, which will in turn shape international responses by local populations and their communities, civilian aid and development organizations, and military forces. Building upon earlier chapters, current research and analysis it defines trans-border regions of violence that can be supported and shaped by long simmering conflicts, forgotten trouble spots and rapid urbanization. It describes the potential for accelerated destabilization of these fragile areas through a random combination of natural disasters, environment and climate change, complex infectious health crises and pandemics, and the merging of natural disasters with technological developments to impact critical infrastructure and systems. These unpredictable combinations can be used by non-state actors and criminal networks to take advantage of these situations to synchronize and/or propel regional conflicts onto the global stage for political or financial gain. This chapter concludes by outlining challenges that these complex risks present to civilian organizations and military forces and describing how these new war scenarios will demand resetting attitudes and developing a new more resilient civil–military mindset that will support increased bottom-up development of aid, development, and security policies by affected individuals and communities described as in Chapter 7 and renegotiating more effective civil–military models discussed in the final Chapter 8.

Cross-border violence

There is very little contemporary violence contained within international borders. Instead it masses within regions that threaten to engulf more territory if the right combination of timing and circumstances converge. These regions of violence can be defined as geographic territory that crosses legal borders. They can be controlled by governments and/or armed groups who almost always target civilians and govern by fear. Their borders may be diplomatically recognized as is the case with Syria and Iraq but in reality they can also be redrawn and redefined by the group in power. This has been the case with the newly formed caliphate combining northern Syria with northwestern Iraq declared by the Islamic State (IS, formerly ISIL or ISIS). Boko Haram, a group aspiring to establish a similar territorial caliphate in northern Nigeria and neighboring countries proclaimed their allegiance to IS in March 2015. Announced via social media this statement is as strategic and purposeful as any formal diplomatic maneuver made by an established nation state. It is meant to convey an image of Boko Haram's rising regional and global power, an illusion of legitimacy, and to use the reputation of IS to spread fear among their enemies (Wooldridge 2015). Both IS, Boko Haram, and other emerging terrorist and insurgent groups are seeking to establish themselves as legitimate states. This raises two questions for the international community. What constitutes a nation state? What role can violence play in shaping it? While these questions often elicit exhaustive and elusive discussions they are worth briefly considering in this context.

The Montevideo Convention attempted to define a state in the 1930s by listing four requirements: a permanent population; a defined territory; a government; and the capacity to enter into relations with other states. Strictly speaking the self-declared Caliphate of the Islamic State in Syria and Iraq fits that description; however, the international community has added a moral and political dimension to this definition that decrees violence and terror cannot be rewarded. This ignores the fact that North Korea, which regularly targets its civilians, is a full member of the UN and that Taiwan, which has treated its citizens well under peacefully elected governments, is not a UN member having been sidelined due to its longstanding dispute with mainland China and is only formally recognized by a few countries (Boyle 2015).

Joe Boyles compares three contemporary strategies in which regions can officially become states in his BBC article "Islamic State and the idea of statehood." One is to get UN membership, which requires the support of the Security Council and two-thirds of the General Assembly. If attempts to achieve UN membership fail then the next best option is to gain official recognition by as many other countries as possible while functioning as a de facto state. Regions that aspire to be states can begin by initiating trading and developing business deals with other states with which they have formed alliances. If successful, they may receive official recognition or that type of approval may no longer be necessary if they have highly developed unofficial networks (Boyle 2015). He maintains that there is no universal definition of a state demonstrated by three simultaneous and differing interpretations of statehood that have emerged from the Syria, Iraq and IS situation in 2015. One is where Syria as a UN member with internationally recognized borders, partial control of land, has a government disliked by western powers who have appeared to bypass Syria's right to protect its own territory. The other is Iraq, also a UN member with defined borders but with only partial control of their territory. It enjoys both political and military support of the West, especially the US, who having recently ended the long and mostly unsuccessful Iraq War began resending military advisors joined by other NATO allies after IS overran Mosul and other areas in northwestern Iraq. The third major interpretation reinforced by combining captured sophisticated military weapons with medieval era military tactics is IS, which has declared itself a caliphate despite a lack of international recognition of its borders and partial control of its territory with no legal right to protect it (Boyle 2015). These subjective geopolitical interpretations of what constitutes a nation gives armed insurgent and terrorist groups such as IS, Boko Haram, and others who make declarations reinforced by developing governance and other infrastructure in territory they hold a stronger claim that they are being victimized by the international system (Boyle 2015).

Modern violence carried out by IS and Boko Haram is descended from a thousand year lineage that includes war, revolution, and terrorism used to conquer territory, establish statehood, and solidify power. (Ginsberg 2013: B7) In a different century their actions might have been viewed as a legitimate part of the process of establishing colonies and consolidating empires on behalf of powerful distant kingdoms aspiring to be world powers. Indiscriminate violence with little or no regard for what would later be called human rights determined which nations would exist, how their boundaries would be drawn, and which groups would ascend to political power within them. It has been claimed that the composition of the ruling elite within every country has been shaped by some form of violence incorporated into their laws, legal processes and systems that structurally reinforces what has been called "law-making violence" (Walter Benjamin quoted in Ginsberg 2013: B7). Peaceful acquisition of territory and creation of new countries are rare. Most modern western powers have a long legacy of some variation of unrestrained violence that has shaped their borders and directly influenced their institutions. This is borne out by the formation of the United States, one of the world's most successful democracies that emerged from a long and violent history.

Established governments and terrorist groups in contemporary regions of violence such as Syria, Iraq, northern Nigeria are once again using an old strategy that Hugo Slim calls "limitless war" to justify and solidify the dominant group's position. These "winners" view violence as a positive "cleansing and creative" agent of change that displaces or kills all designated "undesirable" individuals and groups while consolidating the victors' political and military power (Slim 2008: 25–8; see Chapter 4). Destroying a population's culture and heritage by attacking their revered historic sites and symbols is intended to extend a limitless war strategy beyond physical killing of designated enemies to eliminating all physical evidence of their identity as a group and civilization in ways that deny their existence for future generations.

In its latest manifestations these acts of killing and desecration are staged to be captured by digital cameras that technology savvy IS members can quickly upload as images and video that are transmitted via social media networks and communication platforms. They have developed a similar strategy for instilling and manipulating fears and anxieties held by individuals and groups in regions of the world they cannot physically invade. Particularly skilled in targeting and provoking reactions from major western powers, especially the US (societies with which a number of IS members are familiar), they have also mastered the art of using these and other images to recruit potential followers inside these countries and to inspire others to carry out internal acts of violence on their behalf. This has extended what can only be described as psychological operations designed to increase fear, anxiety, and feelings of vulnerability among their enemies by raising the possibility of random attacks by IS proxies inside their home countries. Once again IS had discovered another means of extending their reach. IS held the type of momentum that gave them an ephemeral but undeniable psychological military advantage that inspired admirers among similar groups in areas of the Middle East and North Africa including Egypt, Yemen, Libya. Their protégés in Boko Haram, Al Quaeda and other terrorist organizations were taking notes.

The problem for opposing countries and military forces has been the undeniable and unanticipated success of IS, which has created greater problems for their enemies. Once ignited violence flowed into vulnerable communities like leaking water and spread like burning oil. A clear demonstration for a classic case study on "new" war scenarios these strategies and tactics are intended to synchronize local and international political and criminal violence, infiltrate local conflicts, social protests, gang violence, and organized crime and terrorism networks to cross and minimize borders while simultaneously threatening a range of countries with similar or different incomes, identities, religions, and ideologies (Kaldor 2012; World Development Report 2011).

The irony is that many of these operations are rooted in textbook military strategies and tactics as basic as divide and conquer. Aside from acquiring US weapons after overrunning Mosul, IS demonstrated the type of clear cut military victory that can be established by combining the right technological skills with sophisticated information and psychological operations and military tactics (in the case of IS borrowed from earlier centuries) that the US and international military forces had never managed to accomplish after over a decade in Afghanistan and Iraq (TRADOC Pamphlet 525-3-1 2014: 14).

In September 2014 then US Secretary of Defense Chuck Hagel stated that US enemies were also developing anti-ship, anti-air, counter-space, cyber, electronic warfare, and special operations capabilities with goals to counter and neutralize traditional US military advantages, especially the "Army's ability to project power" (TRADOC Pamphlet 525-3-1 2014: 49, 11). By early 2015 US Army Chief of Staff, Ray Odierno acknowledged the success of IS (aka ISIL) while discussing a new army operating concept stating that while "our enemy creates multiple dilemmas for us, we don't create multiple dilemmas for them" (Vergun 2015). A 2014 US Army Training and Doctrine Command (TRADOC) paper "Win in a Complex World, 2020–2040" outlined these challenges and the success of IS (aka ISIL) in using a range of strategies and tactics to create and dominate their operating environment (TRADOC Pamphlet 525-3-1 2014). It described how IS used a strategy of murder and brutality against civilians while forming alliances of convenience to organize people, money, and weapons that allowed them to take advantage of the opportunities offered by regional conflicts and weak governance to consolidate their gains and to further intimidate and marginalize competing insurgent groups. By declaring a caliphate IS (ISIL) provided their followers with a type of sanctuary and "strategic depth" that would allow them to launch a broader regional and ultimately global campaign that reinforced their image of invincibility(TRADOC Pamphlet 525-3-1 2014: 14). Terrorist and insurgent groups can easily access weapons equal to those used by international military forces.

For the first time the new army operating concept focused on all three levels of fighting wars—strategic, operational, and tactical, while declaring that the future operating environment is composed of a series of unknowns. The enemy is unknown, the location is unknown and the answer to questions of who, what, where of involved coalitions is unknown (TRADOC Pamphlet 525-3-1 2014: iii). While indicating the necessity for land forces to defeat determined enemies who operate among civilian populations they control, the US Army acknowledged that the challenges presented by "ISIL" also highlighted the need to "extend efforts" beyond physical battlegrounds to "other contested spaces such as public perception and political subversion" (TRADOC Pamphlet 525-3-1 2014: 14).

A growing challenge for international civilian humanitarian, aid, development organizations and military forces is to successfully identify areas of chronic and potential destabilization where the "right" circumstances can rapidly transform a random combination of risk factors, vulnerabilities, and catalytic events into "unexpected synergies" between otherwise independent risks that evolve into larger cross-border regions of violence that support the formation of "new" war scenarios and magnify their consequences (Gamper 2014: 4). This requires a deeper understanding of the characteristics of contemporary complex, interconnected risks.

Understanding complex risks

Complex risks, also known as compound risks, interdependent or interconnected risks, and hyper-risks, can be generally described as a rapid set of events with severely disruptive consequences that cross administrative and national borders with the ability to produce ripple effects on both local and global infrastructure networks, economic sectors, and international civil–military operations (OECD in Gamper 2014: 4). Compound disasters have been described as events causing extensive damage that multiplies the effect of the original cause to prolong recovery. Hyper-risks describe an event or process that triggers another single or series of unpredictable effects and events with the potential to cross borders. All of these are usually shaped by a rapid clustering of what previously appeared to be mutually exclusive causes that combine to trigger "cascading" or domino effects, also described as cascading disasters and failures in infrastructure systems and interdependent networks. In contrast to traditional risk management techniques where assessments have focused separately on each risk, these encompass a range of natural, social, technological, natural and human-caused risks that merge a combination of "drivers" that overlap and intersect to create a larger disaster (Gamper 2014: 5). These drivers include forgotten crises that form areas of chronic violence, urbanization, complex infectious diseases, environment, and the impact of technological developments on critical infrastructure and systemic risks (Gamper 2014).

Forgotten crises

Identifying potential regions of future violence requires that international focus be redirected away from highly publicized conflicts toward quieter but no less potentially violent areas. Many of these regions have been relegated to a lower priority status by the international community when compared to highly visible complex crises and emergencies that demand an immediate response. However recent history has clearly demonstrated how quickly the "right" set of circumstances can merge to link and interact with international causes to suddenly erupt into large scale regional violence. Sidelined by the international community these "forgotten," overlooked potential "trouble" areas experiencing unresolved, chronic local conflicts offer opportunities for insurgents, terrorists, and criminal networks to hone their skills, develop financial links, and establish human and arms trafficking networks away from public scrutiny (UN Department of Economic and Social Affairs 2014: 7).

The 2014 European Commission's Department of Humanitarian Aid and Civil Protection (ECHO) Forgotten Crisis Index regularly publishes a list of indicators for future regional problems. This index has consistently demonstrated a large disparity between countries receiving a high level of international attention and the location of countries that have repeatedly appeared on the index since 2004. The Central African Republic with an ongoing armed conflict that has caused a range of humanitarian crises, Algeria which is home to the longstanding Sahrawi Crisis, and Columbia with its long term unresolved and continuing armed conflict are all top of the index. Other countries appearing on the index six to nine times within the previous 10 years include Nepal, which has Bhutanese refugee issues, Bangladesh with Chittagong Hill Tracts and Rohingyas disputes, Mynamar with conflict in the Rakhine and Kachin States and Myanmar refugees in Thailand, Thailand in the Myanmar border area, India in the Naxalite affected regions, Jamma and Kashmir and the North East India conflicts, and Yemen (Global Humanitarian Assistance Team 2014: Annex I).

Those that have appeared one to three times in the same decade are Georgia, Abkhazia, the Russian Federation in Chechnya, Sudan with humanitarian crises caused by armed conflict, the western Sahara on the Algeria border, Haiti, Guinea, Chad, Uganda with ongoing armed conflict, and the Democratic Republic of Congo affected by conflict caused humanitarian crises. Somalia, Kenya which has a Somali refugee crisis, Tanzania, Sri Lanka undergoing the impact of returning internally displaced persons, Indonesia, the Philippines with its longstanding Mindanao insurgency and Papua New Guinea are also included on the list (Global Humanitarian Assistance Team 2014: Annex I).

Placed on ECHO's high risk list for most of the last decade Yemen began fragmenting into a highly unstable phase in late January 2015 when western-backed President Abed-Rabbo Mansour Hadi and the Prime Minister resigned following the capture of the presidential palace and control of the capital by the militarily dominant Huthi fighters. This was triggered by a dispute between the Huthis, a mostly Zaydi/Shiite movement also known as Ansar Allah, and the President over a constitution that divided the country into six federal regions and used language that the Huthis found objectionable. They suspected the President's advisor, who they had kidnapped earlier in January, of trying to push through the constitution without their consent. While there were few external actors other than Saudi Arabia and Iran to influence the outcome, Yemen's location is in proximity to Al-Qaeda in the Arabian Peninsula and there is the potential for internal violent opposition from the Shafai (Sunni) areas and southern separatists. Al-Qaeda in the Arabian Peninsula is the group that claimed responsibility for the 7th January attacks on the satirical magazine Charlie Hebdo in Paris and Yemen's police academy, which was bombed the same day. Yemenis have been faced with the choice of either compromising and forming an inclusive political settlement or undergoing a violent break up of the country, which would mirror a Libyan type of scenario. The longer the crisis continues the greater the potential for external interference, both positive and negative (ICG 2015b: Syria Calling; ICG 2015c: Yemen Conflict Alert).

In 2015 the International Crisis Group (ICG) identified Central Asia, including Kazakhstan, Kyrgyzstan, Tajikistan, Turkmenistan and Uzbekistan as a region with potential for future destabilization. This is linked to the growing numbers of men and women travelling to the Middle East from countries in the region to support and fight with IS. ICG estimates between 2,000 and 4,000 joined in the 3 years prior to 2015 for a range of political, social, and economic reasons, attracted by the prospect of combat experience and a more devout religious life. If significant numbers of these IS recruits return to Central Asia they have the potential to destabilize not only their own countries but the entire region, which shares borders with Russia, Afghanistan, Iran, and China. Ethnic Uzbeks, including Uzbekistan citizens make up the greatest numbers of IS recruits and are primarily coming from southern Kyrgyzstan, a country that borders Afghanistan, where the Uzbek community was alienated by targeted ethnic violence in Osh in 2010. Kyrgyz, Kazakhs, Turkmen, and Tajiks recruits are also numerous. They are all coming from secular countries plagued by poor governance, corruption, and criminal networks. None of these nations have the resources or skills to cope with the expected return of battle hardened and politically experienced radical Islamists. Instead of attempting to promote religious freedom, safeguard secular institutions, and support European style attempts to rehabilitate jihadis these countries have reacted to these potential threats by implementing laws that increase restriction of religion and sending badly trained police who enforce them by carrying out heavy handed crackdowns that further exacerbate the problem (ICG 2015b, Syria Calling: 1–2).

Urbanization

Global urbanization, increasing at a pace that threatens to overwhelm local and national infrastructure, is another important potential contributor to creating regions of violence as socioeconomic inequality rises both inside cities and in surrounding areas (UN Department of Economic and Social Affairs 2014: 7). Due to unprecedented urbanization over the past 60 years more people now live in urban than in rural areas. This is in stark contrast to 1950 when over 70 percent of the world's population lived in rural areas. Most of the world's largest urban areas were previously located in the more developed regions, but large cities are now clustered in the less developed "global South." In 2014 54 percent of the global population was urbanized, with expectations that will increase to 66 percent by 2050. India, China, and Nigeria are expected to account for 37 percent of the global urban population between 2014 and 2050 (UN Department of Economic and Social Affairs 2014: 1, 7). The potential for these regions to evolve and connect into regions of violence is high.

Much of the world's future population growth is projected to occur in highly vulnerable, less developed countries where many of these dense urban areas are home to densely concentrated and interdependent populations, buildings and services. They are often located along coastal areas threatened by climate change and weather-related emergencies. Many individuals who voluntarily relocated to cities in search of employment find that the only available housing they can afford is structures built on illegal sites from discarded materials and mud without access to public services. Making up a large proportion of growing cities these impoverished and over-populated sites have been generally ignored by local governments who have not integrated them into their planning or public infrastructure strategies. Inadequate construction materials, garbage disposal and sewage systems increase the potential for regional destabilization by leaving them vulnerable to growing pressures on the availability and safety of food supplies, water, energy sources, and the possibility of new health-related risks. With no growth restrictions these urban centers often expand onto lands prone to geographic hazards such as landslides and environmental problems such as toxic waste disposal sites. Increased flows of legal and illegal migrants fleeing other areas, along with internally displaced populations and refugees fleeing conflicts, accelerate this undisciplined urban growth that increases the likelihood that these cities will become incubators for disease and violence (Gamper 2014: 8–9; Patel and Burke 2009: 741; Mancini 2014). This has resulted in a chaotic and disorganized urban landscape where expectations for a better quality of life have not been met.

International concerns about destabilization in North and South East Asia have increased as these areas become more vulnerable to combined risks of urbanization, new and rapidly transmitted diseases, and climate related disasters that put large segments of their population at risk. Asia, likely to be home to fifteen of the top global cities by 2070, is especially at risk from future (2030–40) coastal and urban flooding that would cause widespread destruction to infrastructure, livelihoods, and settlements (Peters 2014: 2–3). Countries at greatest risk in this region include Lao PDR, Myanmar, Thailand, Vietnam, China, Japan, Philippines, Indonesia, Cambodia, Malaysia, and Timor-Leste. It is worth noting that more than 71 million people were displaced in North and South East Asia between 2008 and 2012, and that a number of these countries were listed on the Forgotten Crisis Index, with several others experiencing recent violent conflicts (Peters 2014: 9, 4).

Similar to forgotten crises these areas of chaotic urbanization are ideal for incubating illegal gangs and activities that provide safe havens and support home base environments for dangerous criminals, terrorists, and insurgents who can easily operate undetected by the police and military forces. If local and national governments continue to fail to provide adequate security, employment, infrastructure and services to their urban constituents they leave these populations open to recruitment and exploitation by these armed groups. Cities have also become ideal stages for carrying out politically motivated attacks against innocent civilians. The 2013 Boston Bombing and the 2014 attacks on coffee shop customers in Sydney, Australia are two examples. US and international military forces are especially concerned about terrorists and other criminal groups using uncontrolled urbanized environments to launch long range missiles and other powerful weapons that could threaten external populations (TRADOC Pamphlet 525-3-1 2014: 12).

To successfully meet the challenge of future complex crises the international civilian organizations and military forces will need to develop a deeper understanding of the ways in which urbanization creates its own urban "fabric," ecosystem, climate, society, culture, economy, and governance systems. These urban systems influence and interact within their own geographic boundaries of urban territory, sociocultural activities, and politics to assume characteristics of mega city-states, that have more in common with other giant urban metropolises than with the national governance structure of the country where they are located (Warmsler and Brink 2014, 11). The United Nations has stated that the battle for future sustainable development will be won or lost in urban areas. Their projects are beginning to focus on cities, especially those located in the lower-middle-income countries experiencing rapid urbanization. City and regional governments will need to review their governance strategies to integrate human, national and international security into their policies (Patel and Burke 2009, 741: Mancini 2014).

Successfully meeting challenges posed by these urban areas requires developing better strategies to identify and anticipate the evolution of complex risks specifically linked to these urban areas that are independent from the countries in which they are located (Warmsler and Brink 2014: 11). Operating in congested and restrictive urban terrain will require that international military forces understand the specific political, geographic, technological, and military challenges posed by these and similar urban environments that are either located within or have themselves turned into regions of violence. Both civilian and military responders will need to develop broader more flexible urban focused approaches to humanitarian aid, development and security to be effective when conflicts and emergencies occur in these environments (TRADOC Pamphlet 525-3-1 2014: 12; UN Department of Economic and Social Affairs 2014: 1).

Complex, infectious diseases

The destabilizing potential of infectious disease on vulnerable regions that have recently emerged from long term conflict was clearly demonstrated during the 2014 Ebola crisis. First the crisis began in Guinea, Liberia, and Sierra

Leone, countries that had only recently emerged from long term regions of violence and conflict, which had destroyed their basic health infrastructure. This left these countries with a ratio of one or two doctors per 100,000 people (compared to Spain with 370 and US with 245 doctors) and a young adult population who had little or no education or knowledge about how diseases spread. Initially misdiagnosed as cholera then Lassa fever, the Ebola virus was not correctly identified until it had been circulating in Guinea for 3 months. Once it had taken root it decimated the already small number of health care workers in these countries. Having been considered a rare disease unlikely to appear in well developed countries there were no vaccines and the only controls, early detection, isolation, infection control and quarantine dated back to the Middle Ages. By the beginning of 2015 almost 24,000 cases and 10,000 deaths were reported inside Guinea, Liberia, and Sierra Leone. (Chan 2015; GHA Crisis Briefing 2015).

The drivers behind the spread of infectious diseases share characteristics similar to complex risks in urban environments but they are further complicated by the rapid pace of global mobility and transport networks that extend their impact throughout interconnected systems (Gamper 2014: 14–15). Warning flags and scares about the potential catastrophic effects caused by large scale pandemics had been raised for years prior to the Ebola crisis, but as long as diseases were treatable, eventually contained and for the most part had little or no effect on populations living inside western developed countries few outside of the health field took serious note. Earlier outbreaks such as Dengue and Chikungunya fever in the 1990s had been spread by the global trade in used tires but more recently diseases have spread faster and farther through interlinking networks of tourism, energy, transportation and agriculture that have resulted in a wider range of international consequences. The world health community made pre-Ebola assumptions that "exotic" pathogens, which caused problems in the developing world, would not be a threat to wealthier more developed countries with their higher standards of living and sophisticated health systems. This theory was seriously challenged by the 2002 and 2003 severe acute respiratory syndrome outbreak that began in wildlife markets and restaurants in southern China and quickly spread by air travel to urban areas in Hong Kong, Vietnam, Taiwan, Singapore, Toronto, Thailand, and the US. The international health community also erroneously assumed that future health threats could be predicted. While many were focused on the 2009 H5N1 avian influenza pandemic in Asia along with African forests and Asian cities as the most likely sources of the next pandemic, the 2012 Middle East respiratory syndrome suddenly emerged from the arid desert environment of Saudi Arabia where camels, not chickens were the source of disease (Gamper 2014: 14–15; Chan 2015). The appearance, spread, and adaptability of the Ebola virus in 2014 broke through all previous assumptions and rules.

Three key lessons learned by the international health community about stopping the spread of global health risks following the Ebola outbreak were outlined by the World Health Organization's Director Dr Margaret Chan in a March 2015 speech to the London School of Hygiene and Tropical Medicine. The first is that competent functioning health systems need to be in place before a health crisis occurs. This requires developing a system that is designed to identify early warning signs of unusual disease events, establishing response teams that can track and investigate cases, and designating capable laboratory services to support these outbreaks. Another lesson has been to expect the unexpected (Chan 2015). Do not assume a virus will behave in the same way as previous outbreaks when introduced into a new setting. Ebola emerged at a time when the WHO and other global health experts were focused on acute respiratory infections as the leading global cause of sickness and death (WHO 2014). While Ebola had been a known entity for 40 years it mutated in an entirely new way during 2014. The spread of the disease also highlighted the important role of community engagement and the necessity of establishing trust while seeking their cooperation. Distrust of authorities prompted many communities to hide patients, conduct secret and unsafe burials, and refuse to cooperate with those attempting to track the virus. In 2015 the World Health Organization collaborated with the World Food Programme to bring WHO teams of social anthropologists and epidemiologists to isolated villages to assist in establishing trust with communities in an effort to track down the remaining Ebola patients until all cases were resolved (WHO/WFP 2015). A third lesson is the importance of creating incentives for research and development of medical vaccines and other interventions for diseases mostly associated with impoverished populations. Ebola demonstrated how easily and quickly a sidelined disease could jump socioeconomic groups and global barriers (Chan 2015). More public health spending on general health care, vaccines, antivirals, and antibiotics has the potential to reverse overall vulnerability to pandemics (Gamper 2014: 14–15).

In many ways Ebola was an international wake-up call about the destabilizing potential of infectious diseases that begin in distant locations. It points out the dangers of weak health infrastructure in areas recently emerging from or currently experiencing long term chronic violence and their connection to global systems. It also highlights the link between environment and the spread of infectious diseases. Changes in the global environment and land use for agriculture, irrigation, hunting, and deforestation have directly contributed to an increase in the spread of zoonotics (animal diseases that can be transmitted to humans), food, and water-borne diseases. Migration and climate change have eased the transmission of vector-borne diseases such as malaria, transmitted by mosquitoes. Changes in social and demographic factors such as aging, migration, unemployment, displacement can also contribute to disease outbreaks. These risks are further aggravated in areas where there are persistent socioeconomic inequalities, where resources are lacking and where disasters and conflicts have occurred (Gamper 2014: 14–15). Developing integrated global strategies and interventions to identify and counter future outbreaks of infectious disease and having functional health systems in place that can withstand future shocks whether from climate change or unchecked viruses is the only way they will be slowed or stopped (Chan 2015).

Environment and climate change

Environment and climate change are widely considered to be interwoven into and underpinning almost all complex risks. While the impact of the environment and climate change on reshaping everyday lives has been widely discussed in the context of disastrous weather events and forced migration of populations due to degradation of land and availability of water, it is more recently considered to be the factor driving an increased frequency and intensity of future complex risks.

Increased rainfall causes flood; accelerating wind speeds can cause hurricanes, cyclones, tornadoes and similar destructive storms. More frequent and longer periods of warm weather can lead to intense heat waves and droughts that are associated with a rise in sea levels, and rapid melting of glaciers and permafrost that destabilize hillsides and increase coastal flooding. Flood risk has been estimated to be the greatest natural catastrophic risk to the largest number of global inhabitants as most of the world's cities are located on coastlines, rivers, and other bodies of water. This includes world financial centers like London and New York (Gamper 2014: 14–15).

Climate change has been progressively pressuring cities into a vicious cycle of cause and effect between themselves and complex disasters. Massive urban areas have been increasingly viewed not only at risk to climate change but also the cause of additional climate-related risks through interconnected systems that combine geographic and spatial, environmental, sociocultural, economic and political institutions. These include urban sprawl, lack and inefficient delivery of services, unsafe construction, traffic, paving over earth and overcrowded informal settlements. Sustainability of urban environments is challenged by a lack of green areas, biodiversity, pollution, scarce water sources, waste and wastewater contamination, and radiation. Sociocultural challenges include a lack of security, segregation, aging populations, loss of cultural and historic heritage, unequal access to services, disease, traffic accidents. Economic issues are poverty, inflation, unemployment. Political challenges are a lack of access to political power that controls economic, social, and legal institutions that shape policy affecting urban populations.

This sets up a feedback loop that is caused by the impact of climate change on shaping climate-related disasters, the influence of the resulting disasters on climate change, the impact of inadequate urban development on climate change, and the reciprocating impact of climate change on inadequate urban development (Wamsler and Brink 2014: 5, 25, 26).

This indicates that a timely and adequate response to one disaster may be the key to preventing another meaning that the "right" response to each environmental and climate-caused disaster is the key to preventing another and should include actions that diffuse or stop a potential chain of cascading effects. There have been suggestions that the nuclear power plant disaster in Japan that followed the 2011 tsunami triggered by the Great East Japan Earthquake could have been better controlled or contained if there had been a more flexible response that combined experience, better understanding of the range of potential disaster scenarios, and a willingness for responders to anticipate, react, and improvise to the unfolding situation. The type of cross-sector engagement and collaboration that was necessary to address this crisis included disaster management, environment, energy, public health, and local economic, industrial, and international relations. Averting future crises of this type will necessitate that previously specialized agencies broaden their scope to develop multi-sector policies and coordination processes (Gamper 2014: 22).

Impact of technological developments on critical infrastructure and systemic risks

The impact that sophisticated users of information and communication technologies can have on civil–military relationships and the operating environment has been explored in Chapters 3, 4 and 5. Technological improvements in information, communication, space, and transport networks that facilitate global economic development and cooperation can conversely act as drivers of complex risks. Their interconnectedness fostered by the rapid expansion of users and accompanied by dependence on the internet leaves core systems vulnerable to cyber attacks. This poses significant threats to international financial and infrastructure systems as many remain relatively easy to hack and capable of causing widespread damage. However recent international analysis has also focused on ways in which the presence of technology itself can facilitate and accelerate risks (Gamper 2014: 13).

Aside from the potential for inciting violence and spreading panic, an equal if not greater and less predictable threat is posed by natural disasters that trigger technological disasters (also known as NATECHs), as witnessed in the 2011 destruction of the Fukushima nuclear power plant in Japan. When a devastating earthquake hit Japan on 11 March 2011, it caused a massive tsunami that killed more than 15,844 people, destroyed homes and businesses and set off a chain of nuclear accidents that resulted in the meltdown of three nuclear reactors in the Fukushima Daiichi Njuclear Power Plant. This nuclear meltdown resulted in widespread contamination of water, food, plants, animals, and fish on the Japanese mainland up to 200 miles away from the nuclear plant and directly caused over 150,000 people to be displaced. The total economic loss has been estimated at between $250 and $500 billion and seen as responsible for a slowdown or elimination of plans to build new nuclear power plants in many countries that in turn affected the global energy market (Ray-Bennett *et al.* 2014: 14).

This merging of a natural disaster with vulnerabilities in technological systems has given rise to a new concept of hyper-risks and the threats and costs that their interdependent social/ecological/physical/economic/political networks pose to local, national, and global systems. Tied to more than single events these hyper-risks can also be tied to processes that set up conditions for a series of unpredictable events that are likely to cross borders setting off another series of cascading effects (Ray-Bennett *et al.* 2014: 7).

Understanding future challenges

The emphasis throughout this book has been on examining past and recent experiences in order to develop new ways of understanding how civilian organizations and military forces can more effectively respond to future complex crises. This chapter outlines how old challenges have evolved into new risks that merge with "New War" scenarios into processes likely to simultaneously increase unpredictability while multiplying the number of ongoing crises in future operating environments. Managing the consequences of these hyper-risks that evolve into new war environments will require that the international civil–military community acquire the ability to respond to unanticipated consequences from disasters that interact both within and between different types of local and global political, economic, social and other interrelated systems. These wild card scenarios can result from a series of outlier catastrophes that combine natural disasters, market crashes, terrorist events, and the catastrophic failure of complex socio-technical systems to challenge the ability of institutions and organizations to identify and respond to these rapidly forming sets of new vulnerabilities. These layers of overlapping hyper-risks have produced a new group of hybrid, quasi-natural, or NATECH hazards that are challenging the international civil–military community to rethink conventional strategies. This requires developing systems and processes into increasingly complex but streamlined strategies that will more effectively recognize and address random combinations of events that defy conventional thinking and make ordinary disaster countermeasures ineffective (Ray-Bennett *et al.* 2014: 7).

A key to mitigating future complex crises is to develop an international civilian and military system that will facilitate early identification and management of emerging risks by developing processes that can anticipate and respond to risks. Anticipating risks includes identifying, evaluating and prioritizing potential threats and opportunities. Applying conventional logistics, measurement techniques, and governance frameworks will be increasingly inadequate to anticipate and counter fast moving interrelated groupings of familiar and unfamiliar risks. Responding to these risks will require developing a method to assess and respond to threats identified as creating the biggest problems for specific individuals and organizations. Creating conditions that support a system of "opportunity management" that will improve government and community capacity to adapt to changing environments and take competitive advantage of situations as they occur by responding effectively and protecting the public's wellbeing with the overriding goal of avoiding a major failure is considered to be a successful outcome (IGRC 2014: 1, 5).

The international civilian humanitarian, aid, and development community has proposed a new paradigm to support disaster risk reduction strategies that are focused on anticipating and managing hyper-risks and laying preparations for the unpredictable by building organizational resilience that will counter a broad spectrum of hyper-risks and address the potential ramifications of NATECH disasters (Ray-Bennett *et al.* 2014: 8). Developing reflective responses that combine individual, organized, and critical reflections and reflective strategies embedded within an organization's process and structure has been suggested as one strategy that will promote the resilience of an organization (Ray-Bennett *et al.* 2014: 8–9).

Rather than a separate reactive or proactive disaster response, a reflective response calls upon similar and different "communities of practice" embedded within the organization's structure and culture to expand upon their previously independent responses. It combines and incorporates input from a broad range of individuals both within and outside an organization about their understanding and meaning of different approaches to the same issues. The idea is to build upon these relationships and multiple experiences to develop each organization's capacity and continued education on the range of responses to hyper-risks. This builds the resilience of the responding organization and communities most vulnerable to these types of risks (Ray-Bennett *et al.* 2014: 9).

Resilient organizations and communities are capable of learning from their experiences and adapting to change. They do not collapse when facing a series of unpredictable events and learn how to respond to the concept and reality of the potential for cascading disasters merging into a process of hyper-risks. Organizations and communities that build this type of adaptability are in a strong position to withstand not only internal turmoil but be better able to respond and mitigate disaster risks externally especially in communities at the highest risk. "Robust and resilient" organizations and communities are better equipped to prevent or slow down problems that develop in a hyper-connected world (Ray-Bennett *et al.* 2014: 9–10).

Acknowledging the changing operational environment US military forces have outlined their version of developing resilient responses to future operational challenges in their 2014 TRADOC (US Army Training and Doctrine Command) Pamphlet "The Win in a Complex World." Describing anticipated threats and the future operational environment this publication gives an overview of ways in which diverse enemies that include both nation states and non-state actors such as transnational terrorists, insurgents, and criminal organizations are expected to employ "traditional, unconventional, and hybrid" strategies that will threaten US security and interests. These enemies will avoid US long range surveillance and missile launching capabilities to use low tech techniques such as improvised explosive devices and traditional tactics such as dispersing and living among civilian populations to conceal their activities. As new military technologies are more easily transferred military forces will face threats that "emulate" US military capabilities including precision-guided rockets, artillery, mortars, and missiles to give the impression that US military forces are ineffective. Opposing forces also limit US retaliation and proactive tactics as they take cover or launch these strikes among civilian populations. Countries hostile to the US and its allies may attempt to overwhelm US defense systems or prevent their action by imposing a "high cost" to US-led international intervention in a crisis. Many unfriendly governments and non-state actors have access to and the ability to use complex technological communication, defensive, and attack weapons and related systems that can disrupt or nullify the US military's ability to dominate land, air, maritime, space, and cyberspace "domains." Enemies and other adversaries will increase their operations beyond traditional geographically defined "battlefields" to attempt and inspire internal attacks within the US and other countries by using sophisticated communication and information technologies to shape public perceptions while promoting propaganda and disinformation (TRADOC Pamphlet 525-3-1, 2014: 10–11).

TRADOC also lists several characteristics of future operational environments that are anticipated to significantly impact military forces operating on land. These include increased velocity and momentum of human interaction and events driven by social media and the internet and the potential for what is described as "overmatch" by enemies capable of responding with equally sophisticated and technologically capable cyber and conventional weapons as those used by conventional military forces. This will require a new mindset when developing advanced weapon systems as new technologies will need to be accompanied by assumed capabilities of enemies to emulate or prevent their effectiveness and necessitate a combination of technologies and integration of efforts across systems and approaches to create multiple simultaneous challenges and dilemmas for enemies. There is also the potential for nations to lose control over nuclear assets as well as chemical, biological, radiological, and high-yield explosive weapons during fast moving new war hybrid-risk scenarios where violent extremist organizations move in to defeat government defenses to take possession of these weapons and systems. The US Army foresees specially training army forces who will conduct reconnaissance operations to confirm the presence of these weapons, defeat forces that illegally possess them, and secure areas where these weapons are contained until special disarmament teams can arrive to reduce or neutralize their threat. Domination of cyber space and space is anticipated to continue growing in importance for both global and regional competitive governments, non-state actors, criminal organizations, and others who wish to acquire military power to support their objectives. Satellite jamming capabilities could make conventional military technological weapon systems ineffective or redirect long range missiles to different areas (TRADOC Pamphlet 525-3-1, 2014: 10–11).

US military forces have also increased their understanding of the role that violent transnational criminal organizations play in destabilizing entire regions by setting the conditions for creating and perpetuating corrupt and ineffective judicial, policing, and governance systems. Conventional military forces operating in or near these regions, particularly Central America, encounter levels of murders, kidnappings and injuries equal to or greater than the violence experienced by populations from Middle Eastern and other global political insurgencies. The US Army emphasizes the need for US and other regional military special forces who have the capability to understand complex environments and know how to cooperate with multiple partners to counter these types of criminal activities and networks (TRADOC Pamphlet 525-3-1, 2014: 14).

While countering high tech threats is important it is often low tech challenges that have significant potential to undermine the best intentions. The type of indiscriminate threat posed by infectious diseases looms large on the horizon for merging with other risks to create destabilizing regional conditions that could play into the hands of groups with violent intentions. Ebola highlighted the ease with which a health threat from a disease from which there was little or no protection could rapidly spread without timely recognition or acknowledgement of its potential for destroying or destabilizing large regions and threatening many other global locations. This experience has resulted in a call for the world to develop a collective defense against infectious diseases. WHO's Dr Margaret Chan cites the need for developing new thinking and new approaches to addressing health challenges. This will require investing in high performing health systems that integrate public health with primary care. These health systems will need to develop the capacity and financing needed to support responses that can keep up with sudden demands caused by outbreaks of infectious diseases and humanitarian emergencies. Envision an unanticipated natural disaster occurring within an Ebola affected area already plagued by political unrest and violent militias. The international community also needs to create incentives for research and development of new medical remedies and products that will prevent and contain diseases that are associated with poverty affected populations. Allowing an infectious disease to spread by failing to respond to what was perceived as a disease that did not threaten populations in western developed countries allowed Ebola to accelerate into a rapid and growing global threat (Chan 2015).

Developing the resilience and strategies necessary to effectively respond to future fast changing environments will require stronger geopolitical will and a new type of international flexibility that may result in unexpected and new alliances to prevent new regions of extreme violence from forming. Often the dominant force in both numbers and resources the US military has found adjusting to international civilian organizational viewpoints and expectations to be especially difficult. This is likely to change as smaller numbers of international military forces respond to future complex crises. International civilian organizations have also come to accept that it is necessary to negotiate with military forces while simultaneously recognizing the need to develop additional sources of protection for their long term operations when formal security forces withdraw from that area.

As described earlier many of these complex crises will evolve from some combination of new war scenario that has been triggered and facilitated by a random set of disasters merging with other risks to cause a chain of incidents and technological failures that rapidly interact with underlying violence and ongoing conflicts to escalate into large scale instability and destruction. The inability of governments, international civilian humanitarian, aid, development organizations and military forces to respond quickly and effectively to these emergencies will give well positioned insurgent, terrorists, and international criminal networks opportunities to take advantage of the ensuing chaos to take control of large territorial regions that they can govern through the strategic use of violence. But the complex risk scenarios also open up opportunities for individuals to develop systems that keep or take back control of their communities and shape the international actions and programs that directly affect them. The public voice emerging to reverse pervasive top-down development and implementation of international aid and security policies is examined in Chapter 7. This bottom-up approach will present another set of new challenges to civilian organizations and military forces as they learn to build communication channels that will assist future humanitarian response operations to match local needs with distributed aid while protecting local populations and civil–military staff in dangerous environments.

7 New public voices

The majority of civil–military challenges reviewed in earlier chapters have assumed a primarily top-down construction process for development, aid and security policies but it has become clear that opportunities for local populations to articulate their needs and self-organize have dramatically increased in the digital age. As has been proved time and again in the midst of complex emergencies local communities undergoing crises often become their community's "first responders" after an emergency, disaster, or security-related incident. Their actions save more lives than external "second level responders" who arrive later (IFRC 2013: 73). These first responders can interact with the international civil–military community to identify and prioritize and communicate needs from the site.

Since 2010 the use of new communication, information, and web-enabled digital platforms, which gave disaster and violence afflicted populations in Haiti and Libya a public voice have sparked a reversal in the formation of development, humanitarian aid, and security policies that directly affect their planning and implementation. While international civilian development and aid institutions had long discussed the necessity for incorporating civil society into developing and implementing their programs and policies the process is still primarily top down. Military forces engaged in counterinsurgency or other operations in civilian dominated environments have made similar statements about the importance of engaging and building trust with local populations, but this has been carried out as a top-down mission shaped by specific expectations and goals. Contributions made by civil society and the host country were most often considered as late stage inputs into a framework or program that was developed offsite by external civil–military interveners as part of a hierarchal top-down process.

Despite decades of international investment in policies and programs to reduce risks and threats the resulting disproportionate social and economic costs faced by millions of impoverished communities continue to rise. As the conflict, disaster, criminal, and terrorism environments continue to merge international humanitarian and development organizations have taken the lead in incorporating local perceptions into developing, implementing, monitoring, and evaluating projects and programs. Their increased use of face-to-face consultations and information and communication technologies is leading the way in reversing top-down traditions while increasing program effectiveness. International security programs and missions have lagged behind but there has been increasing discussion among international civilian development and aid organizations about seeking and using input from local perceptions of security strategies that keep them safe to implement programs that reinforce these views (see Chapter 4). As these efforts increase it is likely that military programs, missions, and interventions will eventually follow their lead.

This chapter begins by discussing how the views of local populations are being sought and the importance of including the voice of marginalized and vulnerable populations. It continues by reviewing the United Nation's new global initiative to seek face-to-face consultations and online input into developing the post-2015 global development agenda. The rise of the new digital humanitarians and ways they have contributed to accelerating the use of social media and related technologies during disasters and conflicts is discussed. Finally, it examines the role of local populations in developing short and long term security strategies and development and implementation of conventional military programs and missions are increasingly challenged.

It concludes by suggesting how this flipped scenario will shape future civil–military responses to rapidly evolving complex crises and contribute to the evolution of civil–military models discussed in Chapter 8.

Incorporating local voices

Civilian and military initiatives have often implemented programs and missions based on what they assumed to be their greater programmatic knowledge and technical capability relative to the local populations they have been sent to assist. These types of civil–military approaches were implemented in Vietnam (see Chapter 1), Afghanistan, and Iraq (see Chapter 2) and achieved limited results. In the era of relatively contained conflicts it was possible for the international civil–military community to continue making incremental changes as long as development, aid, and security were envisioned as overlapping but separate problems, but this strategy has become far less effective against conflicts characterized by transnational multiple threats (Chapter 6).

As trans-border conflicts and numbers of non-state actors increase local communities identifying and reporting threats are essential to accurately evaluate the effectiveness of civil–military programs and protection strategies. As hybrid conflict scenarios continue to evolve, often becoming one factor in a series of multiple disasters (see Chapter 6) the input of affected populations will become fundamentally important in increasing local capacity and shaping future international civil–military responses and interventions. Looking beyond single events occurring in isolation international projects are establishing new networks to collect views of local populations that will contribute to developing new and more effective long term interventions that will more effectively address complex crises. The Global Network of Civil Society Organisations for Disaster Reduction (GNDR) initiated a global project Views from the Frontline in 2009, which had collected interviews from over 85,000 community members, civil society organizations, and representatives of local governments by early 2015. Seeking individuals' perceptions of both extraordinary and daily threats GNDR's goal was to collate data from their responses that could be analyzed by age, gender, socioeconomic factors to inform development of local, national, and international policies, plans, and projects (GNDR Frontline 2015: 3). This information would assist in monitoring the effectiveness of policies designed to reduce risk while simultaneously tracking and supporting community resilience strategies (see Chapter 6 for overview of complex risks).

Frontline respondents in South America listed seasonal floods, storms, drought, access to health and education, effects of pollution on farming, environment, drinking water, as well as landslides, earthquakes, fires, traffic accidents, disease, unusually cold weather, alcohol, drug abuse and crime as significant threats to their livelihoods and personal safety. Ninety percent of the disasters and threats prioritized by project participants were experienced by these communities as everyday problems and were addressed as one large interrelated risk rather than as separate issues. In other countries many threats were small scale, reoccurring, and the combined results of environmental, economic, social, and political threats that were part of normal life for over 1.5 billion people living with instability, conflict, informal housing, poverty, local climate change and disaster risk. These issues were often ignored and left unidentified, unrecorded, and unaddressed at the national and international level. With few other choices communities often assumed responsibility for developing their own plans for securing and protecting themselves, and their sources of income, and assets. They managed by staying informed about local threats, their potential consequences, and correctly assessing their community's ability and limitations to address them effectively (GNDR Frontline 2015: 5, 7, 15).

Unheard voices

Incorporating the input of local voices means looking beyond the most vocal contributors to seek the input of individuals and groups less commonly heard. Listening to, re-evaluating, and empowering the most vulnerable groups is essential to achieving more effective results from civil–military interventions and programs. These groups include women, children, elders, those who have disabilities, indigenous groups, ethnic, religious minorities and other marginalized groups. Despite access to resources that help manage and mitigate risks caused by complex crises their potential to contribute a range of valuable perspectives and information and assume roles as leaders and decision-makers in building more resilient communities has often been overlooked (GFDRR 2014: 20–1).

Promoting gender equality, defined here as sociocultural expectations and norms associated with being a man or woman, has proved to be especially helpful in developing long term stable and prosperous communities. This is an important issue in the distribution of political power and economic wealth. Removing barriers that prevent women from having the same access as men to education, economic opportunities, and other resources increases productivity for all. Improving women's status also increases the effectiveness of development and aid programs and can improve outcomes for their children. Allowing equal access to social, political, and economic opportunities for both men and women will lead to more inclusive and representative policies—all of which contributes to the perceived legitimacy of formal institutions and maintaining stable environments (GFDRR 2014: 17).

Addressing gender disparity will also help save lives during disasters as women and girls are especially vulnerable. A 2007 study of 141 natural disasters between 1981 and 2002 by Neumayer and Plumper found that when economic and social rights are distributed equally between men and women related deaths are not significantly different between them. This is in contrast to Bangladesh statistics from 1991 Cyclone Gorky (cyclones are also known as hurricanes) where deaths of women outnumbered men 14 to one among the total deaths of 140,000. Much of this was due to roles assigned to women that made them primary caregivers for children, the sick and the elderly, wearing traditional garments such as sarees that could become easily tangled to restrict movement during extreme weather, and social norms and concerns about privacy and safety that prevented women from leaving their homes or staying in cyclone shelters without a male relative. By 2007 only 3000 deaths from a cyclone with a smaller five to one ratio of female to male deaths were recorded after Bangladesh organized cyclone shelters into safe, female only areas and trained women to mobilize and communicate early warning messages to their communities (GFDRR 2014: 17–18, 20).

Recovery efforts following crises have also been unfavorable to women in both distribution of assistance and supervision of programs. There have been problems distributing relief through women, including a woman's name along with a man's on a newly constructed house deed, incorporating women's needs and preferences into house designs, promoting land rights for women, and developing non-traditional skills to earn additional income, and funding women's groups to monitor recovery projects. Including women in grassroots response and recovery efforts significantly increases and improves the resilience of their communities. Their input contributes to better management of risks from disasters and emergencies. They also tend to focus on smaller pragmatic and innovative projects than men such as seeking alternative water supplies or planting new varieties of crops as opposed to male tendencies to focus on larger projects such as expanding agricultural production (GFDRR 2014: 18–19).

As the international community increasingly acknowledges its limitations in accessing individuals and communities experiencing complex emergencies international development and aid discussions have focused on developing resilience in local populations. Resilience is defined here as a multidimensional collective ability for individuals, families, communities, and institutions to anticipate, endure and adapt to catastrophic events and experiences in ways that allow them to maintain normal functions without losing core identities. On an individual level resilience is interpreted as a "normal and common" response to adversity. On a family and community level it is described as a capacity to anticipate, withstand and maintain normal functions after a disaster occurs when assisted by the right "types, timing, and levels of social support." Resilience of larger integrated human and natural environmental systems is defined by the amount of change a system can withstand while still retaining the same level of control over functions and structures, the ability to self-organize, and the capacity to adapt (Almedom and Tumwine 2008). The process of accurately identifying local risks, threats, and vulnerable groups is the first phase in developing resilience responses in local communities.

Incorporating all voices shifts the international development, aid, security program paradigm from a focus on recipients as beneficiaries to one that solicits the input of valued partners who can contribute unique knowledge and insights. It increases community involvement through information sharing, consultations, and participation by all community members in controlling resources and making decisions. With local assistance international civilian organizations and military forces can more effectively adjust their programs and missions to the scale required. Community-driven systems put resources directly into households, communities, and local governments who can create sustainable platforms for making simple adjustments to operational procedures according to specific needs and circumstances demanded by crises. This will allow communities to expand or reduce their capacity to deal with changes in resources and address vulnerability gaps by changing the status quo (GFDRR 2014: 20).

Crowdsourcing development

Crowdsourcing, a popular online mechanism for recruiting ideas, services, people, and money, is not a concept commonly linked to United Nations Development Policy. However it is less surprising when one realizes that the UN was one of the first frontline humanitarian organizations pushed to cope with digital communication flows during the 2010 Haiti earthquake. Acknowledging the under-utilized power of the international public voice in August 2012 UN Secretary General Ban Ki-moon initiated a ground-breaking open, inclusive and transparent grassroots process inviting global input into shaping a post-2015 global development agenda. Seeking to expand upon the list of Millennium Development Goals (MDGs) that had been part of the Millennium Declaration at the beginning of the 21st century 13 years earlier, he combined face-to-face meetings with online forums to seek views from people on local and global issues they thought important to their lives.

Members of the most vulnerable, poverty stricken and marginalized communities rarely asked for their opinions on international issues were identified and included in this process. National discussions led by members of civil society, academia, and other organizations focused on governance, food security, conflict, inequality, health, education, and the environment (Clark 2015; UNDP 2013).

Early international feedback indicated concern over key omissions from the original MDGs that included governance, peace and security, equality, unprecedented demographic changes and insufficient emphasis on environmental sustainability. They expressed concerns on how the choice of wording in the original MDGs such as using the word "hunger" instead of specifying "food security and nutrition" appeared to have slowed progress in identifying the root causes of these problems and in addressing these issues more broadly. Indicators of progress specified for MDGs also proved to be problematic as they did not incorporate many issues already identified as equally important indicators through earlier surveys and participatory research. These included the presence of effective and responsive national institutions, addressing problems of inequality, building inclusive societies and political systems, providing jobs, sustaining well functioning economies, advancing human rights, ensuring freedom from insecurity and violence especially against women, and promoting livelihoods and lifestyles that preserved local and global environments for future generations. The original MDGs also relied heavily on the political and financial aspirations of each government participating in developing and signing the original Millennium Declaration rather than linking those goals to real life problems (Clark 2015; UNDP 2013: The Global Conversation Begins, 3).

Early consultations also indicated that there was broad consensus that the elimination of violence is a fundamental sign of human progress and that peace and security should be integrated with development in the post-2015 agenda. While usually discussed in a national context with activities carried out by criminal networks including trafficking people and drugs, robbery, accompanied by vulnerability to a combination of disasters, conflict, and terrorism violence was viewed as a transnational challenge to which no country was immune. Concerns were expressed about increased tensions caused by large numbers of refugees escaping conflicts such as the one in Syria that are seen to have a destabilizing effect on security and the job market in the host country (UNDP, 2013: 36–7, The Global Conversation Begins).

A final high level meeting focused on the Global Thematic Consultation on Conflict, Violence and Disaster in the Post-2015 Development Agenda was organized by UNDP, the UN Peacebuilding Support Office, the UN International Strategy for Disaster Risk Reduction and the UN Children's Fund, hosted by the Finnish government in Helsinki, Finland, 13 March 2013. It officially acknowledged the mutually reinforcing cycle between conflict, violence and disasters and recommended they be simultaneously addressed in the post-2015 agenda. Noting the difficulty of getting input from countries currently undergoing conflict in the post-2015 process attendees suggested reframing a fourth module of sustainable development as peace, justice and security. In summarizing the final meeting, Finnish Minister for International Development, Heidi Hautala, highlighted the convergence of these issues while emphasizing the necessity for incorporating the causes and consequences of conflict, violence, and disasters into a post-2015 agenda. This included identifying how these issues have the greatest negative impact on existing vulnerable populations and ways in which they could destroy lives and livelihoods and erase any earlier progress made in fulfilling MDGs. Understanding and eliminating development programs that were insensitive to conflict was of particular concern for their potential to combine with violence and disasters to cause negative consequences that could perpetuate and ignite a new cycle of conflict, violence, and disaster (IISD 2013).

Post-apartheid events in South Africa have clearly demonstrated the importance of linking development with security and justice on both the local and global level to achieve stability and sustainable development goals. Considered to be a relatively stable and prosperous country within the African continent, South Africa has become a destination for the second highest number of migrants from other African countries while simultaneously experiencing challenges posed by transnational criminal syndicates in drug trafficking and poaching of marine and wildlife resources. This large scale migration has caused tensions between local populations and the immigrants competing for jobs, goods, commodities, housing, and local political control. Arriving with expectations for a better economic future many migrants have instead experienced police harassment and denial of basic services. Highlighting and exacerbating persistent long term inequality among South Africans this situation has resulted in continued sporadic attacks on foreign migrants and large scale high profile xenophobic violence, most notably in 2008 and again in 2015. Maritime trafficking of drugs and restricted natural resources reinforced by piracy remain mostly unchallenged due to the lack of cohesive regional cooperation among neighboring countries undermined by limited capacity and resources for local law enforcement, limited sharing of information and intelligence, and corruption. In 2014 the NGO Saferworld urged the South African government to utilize the UNDP post-2015 development consultation process to broaden their immediate and sole focus on domestic issues in order to reframe and prioritize links and interaction between these issues and international migration and transnational crime at the national, regional, and global levels (Saferworld 2014: 19–20).

Best practices designed to broaden participation and strengthen participatory monitoring of development processes were highlighted throughout the global consultations. They include a range of successful mechanisms such as Zambia's use of a citizen voice and action model to facilitate dialogue between communities and government to improve services such as health care and education and the use of M-WASH, a mobile and web-based system with access to 1.7 million people to monitor, evaluate, and report on water and sanitation services. Other examples include the Citizen's Evaluation for Good Governance in Albania, which uses a scorecard for social auditing and gender budgeting and Thailand's iMonitor app, which tracks and evaluates delivery of HIV services, allowing users to log "alerts" if antiretroviral medicines and condoms are not available in health centers or to report discrimination against HIV positive people in the workplace (UNDP 2014).

More than five million individuals had responded to an online MY World survey by the end of 2014 to rank priorities for the future. Health, education, and jobs topped the list, followed by honest and transparent government. As Helen Clark, UNDP Administrator, declared in "The Future We Want," her December 2014 lecture in memory of Dag Hammarskjold, the widely admired second UN Secretary General, the importance of inviting a wide range of participants and incorporating accountability into the development and implementation of a new UN development agenda had been broadly acknowledged. During the consultation and survey process it had become clear that people not only wanted to develop priorities for a global agenda but wanted to also oversee their implementation by accessing information and open data that would allow them to monitor its progress and hold their leaders accountable. What had previously been an external top-down process of identifying, administering, and evaluating programs was reversed almost overnight into a full engagement by civil society in ensuring that they received all allocated program funding and benefits (Clark 2014).

UN Secretary General Ban Ki-moon declared that while the final post-2015 agenda would be determined by governments in a final meeting in September 2015 people around the world were demanding a say in decisions that affected their lives. This included members of civil society, the private sector, young boys and girls, women, disabled and indigenous groups who must feel empowered by the future 2015 framework in ways that reinforced its legitimacy, implementation, and monitoring. Acknowledging that reaching a unanimous agreement that contains the same simplicity and focus as the early MDG framework while effectively responding to future challenges of sustainable development will be difficult, the expectations, yearnings, and articulations by global populations of the world they want cannot be ignored (Ki-moon 2013: Foreword III–IV. A Million Voices).

In her paper "Getting Real About Politics" Alina Rocha Menocal cites that one of the most important lessons to emerge in international development thinking and practice in the past 20 years has been that the challenge of development lies less in what needs to be done, whether building schools or delivering vaccines or identifying the right "technical fix," but in the identifying processes that can either facilitate or obstruct change. This requires a fundamental understanding of the political influences that shape formal and informal institutional dynamics and the incentives they support (Menocal 2014: 2–3). The overarching message is that a national and international environment that supports full, active, and meaningful engagement by civil society will be required to monitor and hold programs and governments accountable.

In early 2015 Helen Clark proclaimed that world leaders had an unprecedented opportunity to shift global development onto an "inclusive, sustainable, and resilient" path (Clark 2015). Some of this will be accomplished by enacting legislation. What is clear is that however the post-2015 global development agenda is written and implemented millions of international voices have stated that participation and inclusion supported by increased capacities and partnerships are the goals they desire for future global development, aid, and security policies (UNDP 2014).

Interactive digital humanitarianism

International attempts to respond to local voices signaling their distress while trying to coordinate rescue efforts in 2010 Haiti soon followed by attempts to map real time evolution of violence unfolding in the 2011 Libya uprising marked the spontaneous appearance of the new digital humanitarians. Merging the interface between technology, data, and aiding populations in crisis they began using communication technology devices and platforms to identify, relay and guide aid to those who needed immediate assistance. This interaction between survivor generated information and *ad hoc* global users of information and communication technologies to facilitate their requests have begun to reshape how the international civil–military community plans and responds to disasters, emergencies, and human-caused crises. While participation during the Haiti earthquake, an environmentally caused disaster, was focused on rescuing survivors, the 2011 Libya crisis was the first time many crowdsourced digital volunteers and CrisisMappers had faced the risk of endangering in country participants by publicly posting their contributions in real time updates. As this issue grew individuals supervising digital volunteers during the Libya crisismapping project struggled to implement a basic vetting system to protect sources sending information using social media, information, and communication technologies (Meier 2015e: 125–7). This required developing two crisis maps—one public and the other confidential and developing awareness among the volunteers about possible disinformation, deception, and how communicating certain details could result in unintended consequences (see Chapter 4).

Access to the internet is the only common requirement for these new international digital volunteers who come from a wide range of backgrounds and different levels of technical expertise. Mobilizing to respond to survivors sending messages during crises, they track crowdsourced and other information relevant to or from crisis areas, which is then posted to mobile and web-based platforms to create crisis maps depicting the progression of events in real time. Many digital humanitarians also belong to international technical communities, originally called V&TCs, volunteer and technical communities, motivated by a belief in an open source ethos that supports making information and data widely available. These volunteers stay in touch and deliver information through Skype, instant messaging, and real time collaborative systems such as Google Documents, collaborative online wikis and microtasking platforms. On a personal level they tend to share altruistic beliefs and a desire to make a difference in the real world, seeing this type of volunteerism as an opportunity to improve and share their technical skills while partnering with humanitarian organizations to support a worthy cause (Capelo *et al.* 2012: 7, 9). Using the most expedient, accessible communication technologies and devices they transmit critical information to individuals and organizations best positioned to assist those most in need. Local media plays a major role by widely disseminating accurate life saving information that helps individuals communicate, organize themselves and their communities, and to identify those requiring immediate aid (Meier 2015e: 1–23; Capelo *et al.* 2012: 7, 9).

As noted earlier the 2010 Haiti earthquake, which marked a watershed moment for this new type of digitally driven humanitarian response (see Chapters 3–5), also demonstrated the technological and organizational limitations of the international humanitarian aid community which lagged behind in coping with the inflows of data and the digitally skilled survivors they sought to assist. The United Nations International Strategy for Disaster Reduction publicly acknowledged the important role of these new digital humanitarians in supporting a more "people-centred" involvement in early warning and response systems that would increase local capacity to organize and provide aid in areas where there are known hazards and threats (IFRC 2013:74, 95). These new technological tools and institutions could be used to balance international, national, individual, civil society, and private sector contributions to support stability and development while managing risks in a proactive and systematic way. The 2013 World Development Report recommended a more flexible, focused analysis that incorporated input from a range of participants to identify risks, context, and priorities in each region and country as they occurred. This was viewed as supporting efforts by local populations to develop resilient strategies and take advantage of opportunities that would improve their situation (World Bank 2013: 3).

The US military providing humanitarian assistance during the Haiti earthquake also became interested in the work of the new digital humanitarians. The US Coast Guard and the US 22 Marine Expeditionary Unit situated off the coast of Haiti communicated by Skype to ask permission from the CrisisMappers to use the data generated by their digital contributors. They later credited SMS messaging and the Ushahidi/Haiti Crisis Map with helping them to locate and rescue survivors and identify NGOs working in those areas (Meier 2015e: 12).

The interest in the rise of international digital humanitarianism can trace its origins to the use of crowdsourcing, crisismapping and Twitter to recruit and deploy digital humanitarian volunteers to monitor and respond to requests and information sent by survivors of crises using these and other communication (Skype, SMS messaging). Humanitarian crowdsourcing was initially used for recruiting and organizing a "capable crowd" of virtual volunteers during the Haiti earthquake who would undertake outsourced tasks from a central individual or organization during the crisis. It is a concept that has been growing in popularity since first used for this purpose. A 2012 guidance manual proposed strategies for coordinating these volunteers with humanitarian organizations and decision-makers to develop an operational framework that would produce the highest quality and most relevant information during crises. Key points recommended before making a formal "activation" request was to review the complexity of the information requested, the size of the volunteer crowd sought, the time needed to complete the task, and capabilities needed from volunteers. These considerations included ease of access to and the availability of the requested data and resources, level of subject matter expertise, analysis, language abilities needed, and the ability to team up with a local network. The confidential nature of the crisis and security of the volunteers were especially important (Capelo *et al.* 2012: 12, 16).

Numbers of volunteers sought should be proportional to the volume of data that would be acquired and processed, the urgency of the task, and the amount of organizational oversight and management needed. Members of small teams would allow more personal working relationships to develop with the potential to evolve into more specific training and personalized feedback from the requesting organization while larger teams would have greater capacity to achieve specific types of results. Assignments could include producing and processing data, problem solving, providing access to personal technological resources, and gathering information from social media, which may not have originally been intended for that purpose (Capelo *et al.* 2012: 17). The degree of oversight needed by an organization's management team would be dictated by the length and intensity of the engagement, the type of information required and the limitations of volunteers. Supervision responsibilities could range from weekly check-ins to being on 24 hour call, making it critical that the requesting organization specify estimated timelines and the level of commitment necessary to achieve their goals. Requests for certain types of volunteer expertise such as geospatial abilities and translation skills needed to be clearly stated. Politically sensitive data might require a higher level of confidentiality that made vetting volunteers and smaller numbers of participants necessary (Capelo *et al.* 2012: 17) otherwise much of this information would be tagged with geospatial links to identify real time information on population displacement, violent conflicts, and emergencies (Shanley *et al.* 2013: 866.)

Crisismapping emerged as another interactive platform that could be utilized as a viable disaster response mechanism during the 2010 Haiti earthquake. Refined during the 2011 Libyan Crisis (see Chapter 3) it evolved into an organized initiative led by the International Network of CrisisMappers (www.crisismappers.net) with over 5,000 members. Now regarded as a global "premier hub for crisismapping and humanitarian response" they hold annual conferences that bring together a virtual and attending community of practitioners, scholars, software developers and policy-makers at the "cutting edge" of crisismapping and humanitarian technologies for disaster response (Shanley *et al.* 2013: 866). CrisisMappers integrate mobile and web-based applications, aerial and satellite images, visualization, live simulation, statistical models, crowdsourcing, and geospatial platforms to support early warning systems, increase situational awareness, and synthesize data. They also develop virtually connected technical devices to assess damage following disasters and emergencies, track potential disasters and the spread of infectious diseases (Shanley *et al.* 2013: 866).

Free online platforms such as Google Maps and Google Earth have increased interest in using and mapping geospatial data. Application programming interfaces have integrated with these platforms to develop new adaptations or "mashups" that are easily accessible web-based maps supported by geospatial platforms such as OpenStreetMap, Ushadhidi, and MapStory that allow users to share, create and organize data in real or slightly delayed time (Shanley *et al.* 2013: 866). The UN Office for Outer Space Affairs launched a platform for information gathered from remote sensing and satellite images that CrisisMappers could access for improved disaster risk management and emergency response. Amnesty International and the Satellite Sentinel Project at Harvard established the beginnings of a human rights and human security early warning system by monitoring violence and movements of displaced persons in Dafur and posting updates on twitter (Shanley *et al.* 2013: 866–7).

Crisis Tweets on Twitter, the free open access social network micro-blogging platform, have given a magnified global voice to survivors and aid givers in the form of a 140 character public text (SMS) message to fellow users known as followers who subscribe to the sender's updates. All messages are posted on searchable timelines and when their location or geo-tag system is enabled tweets contribute directly to creating a crisismap that follows events, identifies, and links priority needs and emergencies to their location in real time. While both traditional and Twitter SMS messaging was used in Haiti its use substantially increased during the 2011 Libyan and later crises when it proved to be an incredible resource. Tweets have evolved to include public one-to-one messages, retweets, and hashtags, which act as keywords to consolidate messages tagged to an event or theme. This allows humanitarian responders and digital volunteers to easily identify trending topics and relevant conversations that provide essential information about events unfolding on the ground. In September 2013 Twitter allowed specified public emergency messages to be delivered via SMS to a user's phone (OCHA/012 2014: 4).

Crisis tweets began overwhelming recipients with unfiltered data flows. To cope with this advent of humanitarian related "big data" Arizona State University developed TweetTracker, a program that can extract and integrate relevant data from messages into an open system that allows users to collaboratively track, filter, analyze, visualize and apply incoming information to maps in real time. By integrating these tweets with spatial dynamics this program creates models that depict dynamic and complex processes. Layering on population, social, economic, and environmental data with this information could produce a model to reinforce socioeconomic sustainability and monitor, warn, and track different types of crises. This model may be most helpful when disasters share common characteristics, but is potentially limited by variable human and social factors not be easily captured or depicted by a solely quantitative model (Shanley *et al.* 2013: 868). More recent efforts have attempted to introduce the concept of standardizing hashtags on crisis tweets for more effective use of overwhelming "big data" information flows. A 2014 report by OCHA, the UN Office for the Coordination of Humanitarian Affairs, encouraged the adoption of standard hashtags during a crisis. One hastag category would be the name of a disaster, one would identify public reports, and one would be solely designated for emergency responses. Each hashtag would serve an essential role during crises by facilitating continued flows of information; assisting in the public tracking of need, people, and supplies; and providing a separate platform for notifying monitors where direct assistance was urgently needed (Meier 2014b; OCHA/012 2014: 7). By 2014 both survivors and aid responders had gained greater skill in the use of global technology and organizational methods in recruiting and collaborating with digital volunteers to collect and share information.

Locally led security

Almost universally top-down externally driven civil–military interventions have lagged behind international civilian organizations in soliciting and incorporating the views of local populations into their programs. Similar to earlier development and humanitarian aid practices, security policies have primarily formed outside the countries where they are implemented based on expectations and assumptions of their effectiveness by international civil–military actors. While external peacekeepers, military interveners, and civilian organizations may be effective in improving security in the short term there have been increased concerns about their effectiveness in establishing a long term inclusive peace where "end users" would ideally shape and oversee programming goals and priorities (Denney and Domingo 2014, 6).

Earliest discussions on protecting civilians during crises and disasters assumed that national security would be augmented or solely provided by external militaries whether they were international peacekeepers or international forces (see Chapter 4) but unintended consequences connected to the length of deployments, timing, and use of force have raised questions about the effectiveness of this strategy (see Chapter 5). In the short term international peacekeepers and military forces can create safe spaces and "windows of stability" where national and local level leaders can negotiate and launch security sector reform processes that provide and facilitate a foundation for early improvements in security (Valters *et al.* 2015a: 5). Achieving long term success requires a more nuanced combination of international, national and local level participation and support that addresses and engages with dynamics of local and national domestic politics. The influence of regional politics and security can also undermine and derail a nation's peaceful goals (Valters *et al.* 2015b: 20). This approach challenges the widely implemented civil–military model that supports "liberal peacebuilding" where democracy, some form of economic markets, and rule of law are promoted as foundations for stability and security. Instead there is growing evidence that addressing the realities and key role that domestic politics play in sustaining secure environments may require engaging with a range of formal and informal security actors. This strategy may necessitate some level of acknowledgment by the international community that many modern nations have themselves been shaped by longstanding violent histories (Valters *et al.* 2015a: 5).

Where UN peacekeeping operations and military interventions have been most effective in stopping and deterring violence their success has depended upon a range of factors that include the existence of a "peace to keep," some form of clear agreement or consensus that armed conflict will stop, and a government with the capacity to support a peace process on the national level. Support for a military intervention by a national leader or coalition perceived as legitimate is crucial for its success (Valters *et al.* 2015b: 20). A 2015 Overseas Development Report on Security Progress in Post-conflict Contexts (Valters *et al.* 2015a) cited the perceived credibility and legitimacy of "elite" leaders and their ability to use patrimonial and other networks to buy, incorporate, or strike deals that bring elites and potential spoilers into a peace process as the key factor in developing locally driven long term security. Another was the participation of formal and informal local security forces and providers in resolving and preventing local disputes, which could escalate into full blown violence. Several suggestions to overcome potential tensions created by a combination of short term and long term security approaches included working with but not for elite interests, viewing external support for developing security institutions as one phase within a long term process of change, and initating locally led development programs early in the process (Valters *et al.* 2015a: 5).

Lessons learned from recent reviews of Liberia, East Timor, and other countries emerging from conflict or occupation reinforce these observations. Not understanding or taking into account citizens' expectations and local realities in the earliest phases of "state-building" can lead to a lack of trust in national institutions that undermines their perceived legitimacy. In the rush to rebuild international civilian organizations and military interveners often ignore the historic context and root causes of conflicts and do not take the time to initiate consultation processes with citizens to ask for their opinions or identify key players and their networks in the post-conflict phase. Skipping these steps can result in biased advice to international civil–military interveners that reinforces old systems of inequality and undermines legitimacy of national institutions, stability, and security reforms (Valters *et al.* 2015b: 45; Kirwen 2015).

Combined security and justice programs, widely regarded as essential for establishing stabilization and development that reinforce the formal security sector through highly technical approaches, tend to ignore effective functioning traditional systems viewed as legitimate at the local level. While imperfect there is an opportunity to invest international civil–military time and resources into already accessible and legitimate security and justice processes in the short term while incorporating benefits from both systems into long term improvements. Most importantly, there is evidence that the involvement of local leaders and relationships in formal security programs and strategies ensures effective protection of local populations. Forums where state and non-state representatives, community leaders, and potential spoilers can engage should be established at the beginning of international civil–military interventions to allow participants an opportunity to negotiate basic issues that will help build a long term peaceful environment that prioritizes protection of civilians (Valters *et al.* 2015b: 45; Kirwen 2015).

Choosing short term security strategies effective in reinforcing sustainable security arrangements is critical for long term stability. Short term security solutions that benefit relatively few by encouraging patronage and centralizing political power among key individuals while ignoring the larger population will increase the likelihood of future instability. This should not be viewed as a security first, development second strategy but rather as an interactive process where trade-offs are inevitable. Long term goals can be embedded into short term policies for security, security sector development, and political stability that will keep external civil–military interveners and civilian populations focused on reinforcing the goal of sustainable peace (Valters *et al.* 2015b: 45; Kirwen 2015).

Unconventional solutions that go against best civil–military practice such as conflating military forces and the police to diffuse violence and counter security gaps in the short term are risky but may also be effective in establishing relationships that sustain peace in the long term. While this approach was successful in establishing short term stability in both East Timor and Kosovo, there was no guarantee these short term measures would be reversed again in the long term. International military missions face unconventional challenges where organized crime networks are stronger than domestic law enforcement agencies in post-conflict situations. Addressing this security gap in post-war Bosnia-Herzegovina and Kosovo required international forces to take on the role of law enforcement focused on serious crimes to stabilize the region. Cornelius Friesendorf maintains that while internal security should be the prerogative of the police this type of unconventional "security first" approach where international military forces pursue serious criminals and protect those vulnerable to attack may be necessary to establish the degree of stability necessary for international civilian organizations to work in these regions. As the new wars continue to blur the line between conflict and crime this military policing strategy may be especially welcomed by vulnerable communities. There are many potential problems in this approach especially when blending duties between military forces and domestic police organizations, but the costs of not implementing this security approach may result in long term national and regional destabilization (Friesendorf 2010: 161).

Reality checks

Being informed and knowing how to proceed can be very different from implementing well informed programs and projects during rapidly evolving complex crises. Future civilian and military interventions, programs, and state building efforts need to take into account the move toward recent problem-driven, locally owned, politically informed approaches utilized by international development organizations that focus on identifying and addressing specific problems. It is imperative that they understand the political economy of institutions and individuals working within a specific frame of reference; focus on realistic, achievable goals dictated by timeline and finances; and learn to identify and take advantage of small opportunities that emerge, especially ones that could be expanded into effective large scale programs (Kirwen 2015).

Establishing national security and personal protection will require taking into account how the use of violence is understood by its range of perpetrators and enablers on local, national, regional, and global levels. The international tendency to primarily view local populations as victims of violence needs to be re-evaluated as survival may require a range of tactics that puts that sole categorization into question (see Chapters 4 and 5). Connecting the different levels of participants and their networks and comparing these relationships to data that has been tracked, mapped, and quantitatively modeled on internal variations of violence over time would more accurately frame the context of the conflict and assist in planning future civil–military responses and programs (Carayannis *et al.* 2014: 30). When major security threats are reduced, socioeconomic issues including inequality become increasingly important and can contribute to instability. While maintaining security is a necessary prerequisite to aid and development projects it will interfere with long term sustainable and equitable development initiatives if it overlooks or is too slow to address underlying social and economic issues that could allow grievances to fuel future violence and conflicts into long term security concerns (Valters *et al.* 2015b: 42).

The biggest reality check of all may be the national/regional/global political environment. One of the most important lessons learned in international development over the past 20 years is that while institutions matter the politics that underlie these institutions matter more. This requires that the international civil–military world consider to not only begin thinking more politically but to respond by working "differently." Working differently includes probing questions asked about the process and how changes are made; the role external actors play in supporting those changes; and how these factors shape the kinds of programming, funding, and staffing needed to implement them (Menocal 2014). Correctly assessing and engaging with the political atmosphere in countries where civil–military responders and interveners are operating is essential to success. Geopolitics have proved to be a nearly insurmountable major barrier to protecting civilians in Syria and elsewhere (see Chapter 4) even as locals use social media and other communication technologies to globally communicate their plight.

Recent lessons learned from politically smart locally led development, equally relevant to aid and security programs and projects, found that while aid donors have found it hard to move from thinking politically to working differently, there is evidence to suggest that when they do so results can be significantly improved. Keys to successful outcomes in the seven cases studied by Booth and Unsworth 2014 shared key characteristics including adaptive problem-solving strategies, willingness to engage in phased learning processes and negotiating relationships that focused on identifying common interests. All interventions were politically savvy, well informed, and aware, locally led, and focused on relevant local issues while incorporating local capacity to arrive at sound and feasible solutions. Case studies were located in India, the Philippines, Democratic Republic of Congo, Myanmar/Burma, Nepal, and in implementing the European Unions' forest law enforcement, governance and trade action plan (to reduce illegal logging and trade in lumber by negotiating voluntary partnership agreements between the EU and each timber producing country accompanied by increased EU country law enforcement (Booth and Unsworth 2014: v, 5–6). Cases fell into three broad categories. The first group included a rural livelihoods program in western Odisha, India; two cases of economic reform in the Philippines, a disarmament, demobilization and reintegration program for ex-combatants in the Democratic Republic of Congo. The EU's forest law enforcement, governance and trade action plan to reduce illegal logging addressed a single problem that was large scale, complex, and broadly defined. Programs in Burma and Nepal were locally led and designed to develop adaptive strategies that identified entry points for increasing engagement within their political systems by organizing around locally relevant issues within a challenging political environment (Booth and Unsworth 2014: v).

Lessons learned from all cases showed that donors and their staff can work directly or indirectly through partners to initiate innovative projects and designs if their internal political and bureaucratic organizational environments are supportive. When these conditions existed funding was flexible and available for repeated design and implementation of projects as they arose and strategically allocated to meet financial needs as they occurred. Aid donors made long term commitments and ensured continuity of staff for the duration of these projects. Conversely, the cases underline why significant aspects of current practice undermine iterative, local problem solving. These include a focus on achieving direct, short term results based on project designs that over-specify inputs and expected outputs; pressure to spend that makes relationships with partners aid-centric and allows insufficient time for iterative learning; and squeezes on expenditure deemed 'administrative,' which, when coupled with high staff turnover, impede the acquisition of in-depth political knowledge and the application of skills (Booth and Unsworth 2014: v–vi).

While politically smart, locally led approaches to development and aid are not mainstream practice neither are they "rocket science." Sharing commonalities with universal best practices in policy-making Booth and Unsworth maintain that the rarity of this approach is an indicator of how detached the aid business has become from everyday reality. They recommend that donors take a first step toward changing their practices and procedures, some of which are relatively recent developments that discourage or stop repetitive processes designed to solve problems. This would allow innovation by committed individuals who were given sufficient time to identify and solve problems. Most of all donors need to eliminate oversimplified concepts of "ownership" and "partnership" accompanied by unrealistic assumptions that significant internal changes can be led by outsiders. Results confirm that when funding organizations are able to facilitate and stimulate change and constructive locally led problem solving, combine technical knowledge with politically aware programmatic approaches, and seek opportunities for the best use of funding everyone will achieve successful outcomes (Booth and Unsworth 2014: vi).

In their 2015 "Reality Checklist: 10 Essentials for Impact at the Frontline" the Global Facility for Disaster Reduction and Recovery listed ten ways to engage local civil society for more effective international responses and policies. These include listening to and understanding the experiences of the people at greatest risk while being sensitive to the local real life challenges from insecure, informal, and fragile environments. Ensuring the inclusion of all groups by leaving no one behind especially those most at risk, collaborating across all groups and levels, mobilizing local resources to build upon existing capacities, knowledge, existing resilience strategies are essential. Ensuring coherence across all levels of development, climate change, (humanitarian aid and security) activities, establishing accountability to local community needs, learning lessons from the past while identifying future trends that can inform the effectiveness of "recovery and development" plans, awareness of and protection of ecosystems and most importantly actively working with members of civil society to achieve these goals (GNDR 2015: 7).

These lessons are relevant for reshaping international civil–military models which have often misunderstood their roles in addressing gaps. Military forces undertaking peacekeeping missions and other types of interventions can easily make many of the donor practices described a reality.

Taking next steps together toward building better military programs will require forces to coordinate and build relationships between themselves and local and global civil society and governments. Using a militarily relevant version of the reality checklist they could collaboratively identify locally led security activities to support and partner with while outlining an implementation plan with civilians they seek to protect. By increasing their ability to adapt and respond military forces could create an open interactive environment that would enhance military efforts and protect vulnerable groups and increase stability (see Chapters 3, 4, 5; GFDRR 2014; www.gndr.org/frontline).

8 Negotiating "new" civil–military space

History teaches us that predicting the future is a precarious business as each generation imagines they face a new set of challenges and circumstances. Few foresaw the large scale invasions of Afghanistan or Iraq before 2001 or the revival of a Vietnam era Civil Operations and Revolutionary Development Support (CORDS) style civil–military model that followed (see Chapters 1 and 2). Combinations of old and new issues shaping complex crises can distract international civilian organizations and military forces by appearing to create a "new" civil–military space with unfamiliar mixtures of actors, technologies, environment and climate changes, disasters, emergencies, and conflicts. Few anticipated the rapid and overwhelming impact that information and communication technologies, social media platforms, and armed or unarmed unmanned aerial vehicles (UAVs) commonly known as drones and robots would have on redefining contemporary humanitarian operations, development projects, and civil–military space (Chapters 3, 7).

Played out against a background of fast moving global politics it is easy to overlook the value old insights have to offer for developing solutions that effectively address modern early 21st century issues.

Drawing upon both acknowledged and misunderstood legacies from earlier civil–military roles and relationships discussed in Parts I and II, the underlying dynamics of the "new" wars and participation of "new" public voices described in Chapters 6 and 7, this final chapter describes this "new" civil–military space. Building upon the introduction of information and communication technologies described in Chapters 3–5 it suggests that conflicts, disasters, and emergencies are being reframed into new combinations of humans, cyber, aerial, and robot technologies that respond and interact in unfamiliar ways. It discusses how these technologies combine with the next generation of civilian-driven UAVs, better known as drones, which were once the sole province of the US military. Now widely available and decreasing in cost UAVs are being increasingly utilized by a new generation of digital humanitarians to monitor human rights, environmental changes, conflicts, disasters, and emergencies.

Exploring the emergence of these technologies while simultaneously reinforcing the value of developing and maintaining good human civil–military relationships, this chapter highlights the importance of prioritizing an interactive evolutionary process for policy development over limited production of a fixed policy. Reorganizing civil–military responses to complex crises into short and long term phases is suggested as one way to meet these more sophisticated challenges. Both this chapter and the book conclude by reinforcing the short and long term value of building good channels of communication and better defined roles in unpredictable dangerous environments. As international humanitarian aid and development organizations, military forces, and other members of the civil–military community have discovered, ignoring these lessons from their past can result in unintended consequences that make contemporary problems larger than they might otherwise have been. When the chips are down and a seemingly insurmountable barrier is encountered there is no substitute for developing and maintaining key human relationships that continue to be the key ingredient for pushing through to success.

Mixing humans and technology

Having only begun to cope with data flows from post-2010 information and communications technologies, the international civil–military community now faces the rapid development and use of new technologies, especially UAVs, commonly known as drones, among an expanding range of civilian users. This has been demonstrated by the introduction and growing civilian use of UAVs in the humanitarian assistance and development space. Until recently considered a solely military technological asset, Kristen Bergtora, director of the Norwegian Center for Humanitarian Studies and Advisory Board member for the Humanitarian UAV Network, has pointed out that the mainstreaming and growing role of UAVs or drones in the humanitarian space reflects a technology transfer from the global battlefield where drones, now intrinsic to modern warfare, will increasingly find their place in the civilian humanitarian sphere (Meier 2014b: Rise of Humanitarian Drone).

Their development has been accompanied by corresponding interest by members of the Institute of Electrical and Electronics Engineers (IEEE) who formed the Robotics and Automation Technologies-Special Interest Group on Humanitarian Technology (RAS-SIGHT) in 2012. Their goal is to develop a new generation of humanitarian robots who can assist in rescue, aid, development, and security scenarios both during crises when human access is difficult or impossible and in areas that have limited exposure to or experience in using this type of technology. The combined use of robots alongside social media platforms, information and communication technologies, and UAVs have produced unprecedented flows of big data, which require the use of artificial intelligence (AI) to manage them. Their use is reconfiguring a "new" operational space, which will be characterized by unanticipated mixes of humans and technologies.

Drones for good

The US military's near virtual monopoly on the use of armed and unarmed UAVs in the post-9/11 era has come to an abrupt end. Where only the US, UK, and Israel had previously used armed drones in conflicts, over 85 countries are now estimated to have some degree of armed and unarmed drone capability. A 2011 study found that there were around 680 active drone development programs run by a range of international government companies and research institutes increased from 195 in 2005. In 2013 the US defense consulting firm Teal Group predicted that the global drone market would nearly double over the next decade (New America 2015), but by 2015 these predictions were already appearing to be low estimates as the dramatic and unpredicted use of civilian operated drones grew especially during disasters, emergencies, and complex crises. Recognizing the advantages of their use in inaccessible or extremely dangerous areas UAVs/drones are offering new opportunities for humanitarian responders to improve logistics in delivering assistance and to explore how they can be used to increase organizational effectiveness. Just as drones have become "intrinsic to modern warfare" it appears they have the potential to become equally integrated into future humanitarian responses (Bergtora and Lohne quoted by Meier 2014b: Rise of Humanitarian Drone).

Similar to their use of information and communication technologies the international humanitarian community has approached the use of UAVs for humanitarian work from both a functional and ethical perspective. This new use of UAVs for humanitarian purposes emerged from the 2010 crisis technology revolution that began with simple calls and SMS texting from disaster survivors, which then progressed to the use of social networking applications, platforms, and smartphones. These technologies began creating new pathways for the flow of information from crisis affected populations by allowing individuals to bypass inept, corrupt, or predatory local and national governments to directly communicate with a wider global audience. This allowed internal populations and their external supporters to crowd-source international cyber communities to virtually come to their assistance while simultaneously mobilizing action at the grassroots level (Shanley *et al.* 2013: 866; Ziemke 2014: Chapter 7). As a technical and logistical asset UAVs add to this outreach with their capability to collect more information during disasters, confirm physical damage on the ground, and improve effectiveness for the delivery of humanitarian assistance during crises. As smaller and less expensive UAVs become available their use is expected to increase among other NGOs, large international and civil society organizations, civilian crisis volunteers, and governments (Meier 2014b: Rise of Humanitarian Drone). Similar to the ethical concerns concerning the availability, vetting, and use of information discussed in Chapters 3–5, drones simply increase the flow of unfiltered information. They build additional layers of vulnerabilities through previously unseen photos and huge flows of information with relevant features increasingly identified by humans to shape the development of crisis-specific algorithms generated by AI programs.

Acknowledging that the use of the word "drone" invokes a certain set of lethal associations civilian members of the international humanitarian community have advocated for the use of terminology such as "unarmed unmanned aerial vehicle" or UUAV to clearly distinguish their use in non-lethal operations from the more familiar armed predator drones designed and used by military forces to attack enemies. To further clarify the difference in goals the term UAV is suggested for use when referring to both armed and unarmed vehicles and the word drone only when referring to armed UAVs (Larrauri and Meier 2015: 3).

Challenging early assumptions that only the most technologically savvy users would be capable of utilizing UAVs/drones members of Digital Democracy, a group that promotes co-learning, co-creation, and co-experimentation brought drone parts and glue to remote indigenous Wapichana communities in tropical areas of Guyana, South America. They intended to reconstruct them so that locals could fly them over their villages in an effort to more closely monitor local deforestation and illegal logging in their area. Earlier use of satellite imagery for this purpose had been of limited value due to the low resolution of the photos. Community members, none of whom had prior engineering experience, collaborated to build and fly Digital Democracy's drone and when necessary improvised by using local materials to re-engineer a broken part. Together they collected data, constructed a detail map, and began familiarizing all members of the community teams with flying and smoothly executing the entire process from mission planning to processing images (Meier 2015c: Indigenous Community).

Following the first earthquake in early May 2015 The Humanitarian UAV Network (UAViators) organized in 2013 found themselves providing coordination and oversight to over nine humanitarian UAV teams operating in Nepal. These teams voluntarily liaised with the UAViators to coordinate their work to survey and assess damaged areas (Meier 2015e: Humanitarian UAV Missions). Drawing from Nepal and earlier experiences using UAVs for humanitarian missions after the devastating hurricane in Vanuatu (March–April 2015), the Philippines, and Haiti, Patrick Meier, a leader of the global digital humanitarian movement and the Humanitarian UAV Network compiled a handbook for the use of UAVs/drones during crises. Emphasizing that the list of guidelines was at very best the beginning of a minimum set of best practices, Meier formatted the handbook into an open interactive list of best practices that were organized as an operational checklist divided into pre-flight, in-flight, and post-flight sections. He invited the humanitarian, UAV, and research communities to contribute toward refining and improving the guidelines by directly adding their comments and updates to the handbook posted as an available openly editable Google Doc. Future versions would add a list of best practices for operating UAVs with "transportation payloads" (Meier 2015e: Humanitarian UAV Missions).

The "drones for good" movement represents a technological shift in scale for members of local communities, citizen journalists, advocates for human rights and social change movements. Many of the UAVs flown by these civilians are increasingly small or micro versions of the larger drones making them especially maneuverable and somewhat more difficult to target and destroy when flying. Utilizing UAV capabilities for aerial surveillance provides an additional layer of observation in conjunction with the one on ground level. This expands the concept of public spaces to include less accessible rooftops and courtyards. They allow individuals and organizations to bypass traditional ground or aircraft delivery of humanitarian assistance by sending swarms of UAVs to ferry supplies to civilians trapped in conflict zones.

The increased deployment of UAVs/drones during complex crises has paralleled the international rise in "digital volunteerism" by users of Twitter, Facebook, YouTube, Instagram, Tumblr, geo-tagged information, and additional instant messaging platforms, which has allowed individuals to share real time information about unfolding disasters, conflicts, and other issues with international technical and issue focused communities (Shanley *et al.* 2013: 866; Ziemke 2014: Chapter 7). Following his observations of civil society's new use of UAVs/drones Austin Choi-Fitzpatrick has proposed a framework that guides non-state, non-commercial uses of UAVs. He suggests focusing on "subsidiarity" by proposing that decisions to use drones should be made on the local level and they should only be used in situations when a less sophisticated, invasive, or novel mode of transport is either not available or feasible. Operators need to take special care to ensure that drones do not collide with humans, each other, or other emergency aircraft. He suggests that the rights based on a "do no harm" approach practised in the development and humanitarian aid communities equally applies to drone use and policies for ensuring personal physical and material security should be planned and implemented (Choi-Fitzpatrick quoted by Meier 2015b: Drones for Good).

Questions to ask before operating UAVs include: Do the risks of using UAVs outweigh the anticipated benefits? Is the use of drones in the public interest when they are used for the purpose of investigative journalism and on behalf of the local communities? Awareness of and consideration for privacy issues related to the collected data is a key ethical and security issue. Protecting the raw data collected by these drones is essential as it can be co-opted by governments, opposing insurgent groups, terrorists, and others who attack populations and individuals when their locations are exposed. Data must be secured and protected against physical or digital theft or destruction in all phases of collection (Choi-Fitzpatrick quoted by Meier 2015b: Drones for Good).

Refining the civilian use of UAVs/drones to assist in humanitarian crises is in its earliest formative stages. Based on evolving technologies that make it increasingly easy for civilians located anywhere to construct, replicate, and utilize there is evidence that they are already being integrated into crisis responses in a cost effective and intentionally ethical manner. Similar to the early use of text messages and social media platforms in Haiti, the deployment of UAVs/drones for humanitarian use is pressuring large international civilian humanitarian and development organizations, which have traditionally lagged behind in technological innovation to reframe their understanding and evaluation of a technology that is already being integrated into complex crises responses (Meier 2014b: Rise of Humanitarian Drone). These organizations will be required to develop and establish corresponding communication needs and protocols that will synch them with independent users as UAVs are operated in an "ecosystem of technologies" that necessitate their integration into communication devices such as satellite phones, broadband global area networks or other high speed internet solutions, walkie-talkies, back-up electric generators and recharging systems to make them effective. When these systems are not available UAVs/drones do have the ability to fly to neighboring sites and nearby countries where their data can be more easily uploaded (Meier 2015e: Humanitarian UAV Missions).

Big data and artificial intelligence

The overwhelming information flows created by volunteers, survivors, and impacted populations using earlier social media and communication platforms during crises was initially collated and sorted for relevance by human volunteers. Text-based data now include mainstream news articles, tweets, WhatsApp and other messaging applications. Images are generated by visual applications such as Instagram and professional photos that accompany news articles, images generated by satellites and aerial devices such as UAVs. Videos are broadcast via YouTube, online television channels, Meerkat and similar video-enabled applications and devices. Incoming information from these communication technologies along with the feeds from UAVs and drones are now generating overwhelming amounts of information that has become known as "big data."

These big data flows challenge users to develop systems to identify, collate, vet, and process this information as rapidly as possible in order to provide meaningful and timely assistance during crises. In the earliest days incoming digital information and data from social media, SMS messaging and other platforms were manually sorted and re-released, but this was time consuming and limited to the numbers and capacity of the humans analyzing them. During disasters, emergencies and complex crises the rate of flows as disaster affected communities have become increasingly digitized has created rising demand in the use of artificial intelligence to manage information flows outside the government and military sphere. Manually finding relevant, credible, and actionable text, images and videos among the data generated during major disastrous events and complex crises has become untenable (Meier 2015: Humanitarian UAV Missions).

Meir makes the case that both crowdsourced digital humanitarians and traditional humanitarian organizations are faced with the same choice of either attempting to "continue business as usual" with human-led analysis that results in corresponding limited and eventually falling behind unfolding events or adopting big data solutions along with the rest of the world. He notes that in the future digital humanitarians who focus solely on crowdsourcing to gather and analyze information will be fighting a losing battle as the range and volume of text, images, and video increase and other data types increase (Meier 2015e: Humanitarian UAV Missions).

The big data challenge is being addressed by organizations outside the international humanitarian crisis response sphere that have been using AI in combination with some level of human input to collate and analyze big data. AI utilizes a technique called machine learning, which mirrors human learning. This system initially builds upon human input to develop a program that finds relevant solutions. This interaction between humans and machines interaction has been demonstrated in the experimental use of artificial intelligence for disaster response (AIDR), which utilizes crowdsourced human input to automatically identify relevant tweets and text messages generated during crises. The volunteer digital crowd initially tags tweets and messages they determine to be most relevant to the unfolding crisis. The AI engine then learns to recognize these patterns for determining relevance from real time incoming data. Based on this input it then launches an AIDR system that automatically identifies future tweets and messages. Currently AIDR is one of the few if not the only big data solution that combines crowdsourcing with real time machine learning for disaster response. Huge numbers of crowdsourced international digital humanitarians are needed to tag incoming data as quickly as possible in the earliest stages of crises to get the AIDR system up and running as timelines are a critical factor for effectively responding to disasters. However once an algorithm is created that accurately identifies tweets and messages that contain urgent information and immediate needs, the same algorithm can be reused without having to repeat the process of using human input and crowdsourced volunteers provided the AIDR system is applied to the same type of disaster—weather, environment, conflict, health and any other type for which it has been previously programmed.

Photos can be treated with the same analytical method. Metamind, a Silicon Valley start-up recently developed an AI system that utilizes the same process of depending on human input to initially develop and guide an AI algorithm process that will learn to identify relevant visual clues and then compile and collate pictures that show infrastructure damage following catastrophic events. This same collaborative process between humans and technology is being applied to develop AI algorithms that will detect relevant features in videos, satellite, and aerially produced images. A UK company, WireWax, has been using this approach to automatically identify a range of images in videos, and researchers at Carnegie Melon University are developing AI to detect evidence of gross human rights violations in YouTube videos from Syria (Meier 2015d: Artificial Intelligence).

For the foreseeable future all types of AIDR analysis will continue to be dependent on the quality and range of human input. More sophisticated input and larger numbers of contributors will result in developing algorithms that will produce more effective responses to specific disasters and crises (Meier 2015d: Artificial Intelligence).

Patrick Meier, a leader in the field of digital humanitarianism, has advocated collaboration between humanitarians and tech firms interested in developing AI platforms and programs that will more accurately detect specific types of human rights abuses and visually identify a range of infrastructure damage and resulting emergency human needs. Crowdsourced groups such as MicroMappers are already integrated into this hybrid human technological process as they assist with the initial input of training data for tweets and pictures. This input develops into accurate AIDR algorithms, which can be used to create detailed crisis maps in real time. Unlike private business solutions crowdsourced platforms such as MicroMappers is free and open source and offers the same or similar filters for incoming social media, photos, videos, and mainstream news flows (Meier 2015d: Artificial Intelligence).

Meier predicts that the future of humanitarian information systems will be to choose the most relevant programs in an "Alg" or Algorithm Store rather than an App Store. The user will be offered a choice of algorithms, which have been already "trained" to automatically detect specific types of words, images, features in texts, pictures, and videos generated during different types of disasters. These algorithms will be able to "talk with each other" with built-in capacity to integrate other feeds from real time sensors, satellites and data sources of choice by utilizing existing and developing new data fusion techniques (Meier 2015d: Artificial Intelligence).

Humanitarian robots

The development and use of robots in humanitarian crises has grown dramatically since 2013 when members of the IEEE began seriously focusing their mission on "advancing technology for humanity." Working on a range of projects aimed at improving the conditions of those living in poor and underserved areas they developed technologies specifically tailored for those communities. Forming a separate group called RAS-SIGHT in 2012 the IEEE's goal was to leverage existing and emerging technologies for the benefit of humanity and to increase the quality of life of the underserved. They would collaborate with existing global communities and organizations to share information and align goals. Engineering for Change (E4C) and the American Society of Mechanical Engineers and Engineers without Borders–USA are also members of this global community. Collaborating on a global scale and taking a "bottom-up" approach will help to better identify and understand a community's problems and requirements and ways to apply the technology and expertise that will best address or solve them. The goal for all participants was to provide a meeting place where solution seekers could directly interact with solution providers to support the development of sustainable and accessible technology-based solutions to local problems which could then be continued and maintained by the local community (Pretz 2013).

Working alongside UAVs/drones (which are technically another form of robotic technology) robots and related technologies are becoming powerful tools in disaster mitigation, response, and recovery. While UAVs/drones can fly rapidly over affected areas to survey damage and transmit information after an event such as an earthquake or bomb blast, robots can incorporate that information to operate on the ground in all types of terrain above and below the earth and sea to perform tasks in inaccessible, extremely dangerous, or highly contaminated areas where humans and conventional tools cannot. They can assist in reducing the possibility of unintentional cascading of risks (see Chapter 6) by repairing leaks in underwater oil or damaged nuclear plants. Robots offer a type of emergency response unavailable to humans as they reduce and simplify civilian and military responders' tasks during the different types of changing complex crises described throughout this book. Their ability to integrate rapidly transmitted information flows generated by UAVs and AI programs while simultaneously mapping and accessing highly sensitive and dangerous areas unavailable to humans are predicted to make robots into essential tools by 2025 for all types of complex crises response (UN World Conference on Disaster Risk Reduction 2015).

Reframing international responses

Responding effectively to these new mixes of human actors and range of technologies and avoiding unintended consequences will require international civilian organizations and military forces to reframe their regional and global strategies to build upon relationships that respect differences, acknowledge limitations, and capitalize on strengths. Earlier discussions in Parts I and II that focused on negotiating international civilian and military roles, space, and reframing civil–military models made the case for building trust and developing more adaptive and flexible human relationships and roles in dangerous environments. Part III has described "new" characteristics of recent and anticipated complex risks, disasters, and conflicts and how they are projected to mix with older issues to impact future international civilian and military roles. It suggests that both civilians and military forces operating in the field/theater (i.e. local) and external strategic policy planners (i.e. global) will need to interact on a local/global level to develop more effective civil–military policies and responses.

A combination of new thinking, skills, and old knowledge will be essential to develop adaptive, agile, nuanced responses by members of the international civilian and military community. Assembling a different range of state and non-state civilian and military actors who bring a wider range of perspectives, ideas, creative problem-solving skills to address complex urgent issue will increase the effectiveness of international responses. Beyond reframing international civil–military thinking negotiating this complex space will necessitate that international civilian and military interveners incorporate local voices into their risk strategies and operations while continuing to carefully monitor information and vet sources (Editors 2012: 4). While individual and organizational relationships remain the bedrock for developing effective civil–military responses to crises both must learn how to operate in these environments by addressing and managing the impact of communication, analytical, and physical technologies, which will shape these relationships. Developing more interactive policies that emphasize process over policy and reorganize international civilian and military responders into groups of first and second responders who are more closely aligned to short and long term needs will lay the groundwork for long term sustainable programs and projects.

Focus on global process versus global policy

The concept of a global public policy has existed since the 1980s. It has been described as an area that includes prevention of war and more recently economic, social, and ecological issues, which usually share one or more of the following characteristics. They are trans-border problems such as refugees, pollution, and economic policies that may originate in one country but have ramifications for neighboring areas. There may be a competing conflict between countries over the use of what would be seen as the international commons—the oceans, seabed, outer space, environmental atmosphere, electromagnetic spectrums, and large resource rich geographic locations such as Antarctica. There may also be internal issues and problems shared within and outside national boundaries that are common to many countries, which include a broad spectrum of social, economic, environmental, and political issues such as illiteracy, rapid urbanization, population growth, human rights abuses and destruction of the environment. All these problems are linked to overarching global issues that are beyond the capability of one state to address individually. The interconnected nature of these issues reinforces the need to establish a global policy framework that encourages global cooperation to address shared problems (Miller 2014: 497).

While ideal in concept the reality of developing global policy is often a long slow process that requires many years for a consensus to be reached. However this process is viewed as vital to establishing a social and political space that merges commercial and public domains into a public space that supports discourse for making and implementing polices. While this has the potential to be transformative it can also challenge national autonomy by establishing an independent network that makes decisions and procedures. It can be a space that supports the participation of large established international organizations such as the United Nations or one that focuses more narrowly on single issue agendas that address conflicts, food production, technology, disarmament, and development. These can be special sessions or huge international conferences (Miller 2014: 98).

This approach to making global policy focuses on process. Participants avoid advancing a single national view or approach by examining the issues from a macro international perspective. It reinforces the advantage of mutually identifying and addressing shared cross-border challenges that demand and will only be solved by a cross-border solution. Formal attempts to solve these types of challenges often result in enactment of a type of "soft" law that supports a "hard" law, which reinforces the authority of individual countries and international organizations. While establishing outcomes and a final policy of this type usually involve a very long and convoluted process there has been renewed understanding that the greater international value may lie in experiencing the process itself. These are processes that focus on negotiation, gathering, and dissemination of knowledge rather than solely on achieving finite outcomes. They are continually evolving processes that utilize a partnership approach to support and incorporate dynamics in participant relationships through formal and informal forums. They are issue driven and transcend focus on one state or actor by addressing short and long term issues (Miller 2014: 499).

The vulnerability of this approach is that this process has often been perceived by less politically powerful countries as being driven by the international norms established by the interests and institutions of dominant political world powers. Processes are also challenged by the same type of tensions, chaos, and politics that often troubles other interactions between and within nations. Similar to attempts to establish international law global policy-making struggles with enforcement, compliance, and accountability with these policies. Assumptions that other countries will act rationally, in good faith, or as good international citizens in compliance with these policies are routinely challenged by aberrant or unanticipated national behavior (Miller 2014: 499–500).

However engaging in these processes deepens understanding of ways in which regimes, networks, and partnerships can overcome these problems to cooperate in addressing compelling shared complex issues. Massive forced internal and external population displacement brought about by complex crises and other pressures have reinforced the need for supporting a "living" interactive process that responds to the evolving characteristics of crises and local needs as the most effective strategy for developing formal policies, projects, and programs that will be most effective in responding to current and future complex crises.

James Milner has clearly defined the difference between policy as a product and developing policy as a process in his analysis of global refugee policy. Directly relevant to international civil–military policy formation in complex crises he observes that a range of actors in the international community has tended to focus more on the content of the final product of the policy itself and less on the process of shaping that policy or factors influencing its implementation (Milner 2014: 6). He notes recent work conceptualizing a policy cycle that incorporates four steps. First is setting an agenda where a policy problem is identified as requiring action and a process of policy formation is established to incorporate a range of proposed and considered responses. Decisions are then made to take a chosen course of action. This action results in policies that become operational followed by a policy evaluation process that attempts to measure its effectiveness. This policy cycle framework can be applied on all levels—local, national, regional, and global. It looks beyond the formal final policy to examine the range of factors, interest, and actors that define how that final product was developed, implemented, and evaluated (Milner 2014: 7).

Milner suggests that separating a policy into a product and process raises two separate sets of questions. Examining a formal policy document includes questions such as who are the actors who influenced setting the agenda and subsequent pre-policy discussions? Where are the key decisions being made and how are these decisions influenced by the interests and beliefs of the various actors in formal and informal settings? If policy is approached as a process there is a different emphasis on meaning and impact once it has been created. Rather than focusing on the decision making there is an overall examination of the process from initial development to implementation. How did the meaning of the policy translate from formulation to implementation? What were the factors affecting implementation of the policy and what do those mean in the local, national, regional, and global context (Milner 2014: 14)?

One pressing global problem that is very much in need of an international process to develop a coherent policy has unexpectedly appeared in the millions of refugees and migrants forcibly displaced by conflicts, disasters, and poverty who are seeking better lives. The international community was confronted with this latest challenge when they became bystanders to the seemingly unstoppable Syrian Civil War, the emergence of ISIS (IS/ISIL), and the sudden Ukrainian/Russian crisis. Facilitated by criminal networks trafficking thousands of human migrants and refugees fleeing North Africa, Myanmar, and other crisis afflicted areas this crisis has seemed unstoppable as these syndicates operate with impunity and abandon survivors to their fate inside randomly floating boats or on European and South Pacific shores.

In his March 2015 report Protection in Crisis, Forced Migration and Protection in a Global Era, a project of the Migration Policy Institute, Roger Zetter warns that a wider strategic framework that encompasses migration management, state development, and resettlement support along with traditional asylum needs to be developed. This would provide the basis for a coherent international protection policy that shifts emphasis from the time consuming process of establishing the legal status of right of refugees and access to protection to form policies that immediately address individual and group vulnerabilities regardless of the legality of their entry. He recommends that the current system of different standards and types of rights, opportunities and provision of protection between countries, with northern destinations seen as "premium asylum status," be unified and equally allocated to forced migrants regardless of their geographic location. This would slow mass movements and increased strain on certain countries and parts of refugee system (Zetter 2015: 2).

Achieving this would require that the international community address the increased politicization of protection at a time when focusing on procedures and regulations that attempt to manage the crisis have become increasingly futile. Making protection of forced migrants a political issue has only complicated and increased the necessity for governments and policy-makers to take action on policies that address the increasing needs and gaps in protecting migrants more challenging while doing little to lessen their inevitability (Zetter 2015: 2).

This type of international challenge underscores the importance of prioritizing the process of developing policies over the execution of finite policy products. Politicizing or ignoring a problem of this scope only increases its potential to destabilize entire regions that include the developed countries, which are the migrants' destinations. This issue affects so many different international regions that to view forced migration as less than a global humanitarian and security policy challenge is to misunderstand the seriousness of the problem or the type of cooperative solution that it requires.

Two-phased civil–military responses

Emphasis on responding effectively to future emergencies, disasters, criminal activities, and conflicts has not only focused on improving communication and negotiation of roles between international civilian organizations and military forces but on including additional conversations between these civil–military actors and the populations they are attempting to assist. This three way conversation between civilian organizations, military forces, and local populations they are seeking to help will require intellectual agility that combines an old school analytical depth and ability to analyze complex issues with a new school understanding of concepts and evolving uses of communication applications, AI programs, aviation, and robotic technologies. This new civil–military space includes a combination of actors in the "new" wars (see Chapter 6) and "new" public voices of local populations who seek input into shaping aid, development, and security policies that directly affect them (Chapter 7). While addressing core challenges such as building better international civil–military communication and trust, protecting civilians, and understanding processes of violence will continue to be essential (see Chapters 3–5), the ability to develop effective responses to random often chaotic combinations of humans and technologies will become equally important.

These new local/global conversations and human/technology mixes bring a range of new challenges to international civil–military responders. The word responder is deliberately chosen as the international civilian and military community come to terms with their limitations in responding, containing, or stopping waves of crises caused by disasters, emergencies, conflicts, and the resulting flows of mass human internal and external displacement and migration.

A recent international response to emergencies has been to encourage and build local individual and community resilience (see Chapter 4) but random crises can quickly cascade into a series of wide ranging threats that are beyond the capacity of any single community to control. This calls for reframing international civil–military humanitarian, development, aid and security responses into two phased tiers of first and second responders with different actors and goals and a different type of strategic planning and operations. Managing the new mix of random crises, humans, and technology will require re-evaluating and developing a more sophisticated two phased civil–military response during early and later phases of a wide range of complex crises.

Regrouping international civil–military models into two tiers of first and second responders will be more effective in meeting challenges and taking advantage of opportunities as they emerge. This approach would utilize the best individuals and ratio of civilian organizations and military forces necessary to address short term needs while laying the groundwork for more sustainable programs, institutions, and security. While there would be some overlap first and second responders would have a different mix of civilian and military personnel and goals, and objectives.

First responders

In this phase it is assumed that local communities and individuals will mobilize immediately following disasters, emergencies, and violent events to tend to medical and other survival needs and to inform external civil–military humanitarian responders of events in real time. In this context first responders can be seen first as local individuals and locally dedicated civil–military personnel followed by international civilian organizations, and military forces who fit the more classic concept of interveners. Many of the classic early intervention civilian and military roles described throughout Part II fit this description; however, they would be entering this rescue phase with a finite timeline and a plan for laying the early groundwork for sustainable activities carried out in the second phase. First responders could include civilian humanitarian, aid, and government emergency responders who share space with security forces that include special forces and other military groups responding to disasters and emergencies. Immediate medical, technical, technological, communication, and security needs would be addressed by this combination of local and international civilian and military actors. Ideally civilian organizations and military forces would have aid and security protocols in place to prepare for possible mass displacement and migration out of the affected areas, which have been previously discussed with second phase civil–military responders who would be located both within the crisis area and in adjacent regions.

This space would be shared with all local and externally available communication and information platforms and technologies, AI programs, UAVs or drones, and robots from the very beginning of the crisis. First responders from the formal international civil–military community would be prepared to cope with this information and technology blast by developing operational plans to share real time information with each other and outside activists and volunteers to aid faster recovery of victims during unfolding crises. This would allow all responders to quickly identify and address immediate short term medical emergency, survival, and security needs. These responders would be assigned to crisis locations with specific limited tasks for a finite period of time that would simultaneously respond to emergency needs while laying the groundwork for more sustainable programs, institutions, and security initiatives implemented by second phase civil–military responders. Throughout this process local and international first responders would be communicating with the second phase civil–military responders in an interactive process to begin shaping long term aid, development, and security programs that are designed for specific needs. Local populations would be incorporated into this process from the beginning of an international crisis response.

Second responders

Second phase civil–military responders composed of a different mix of civilian and military personnel already engaged in interactive policy formation with their first responder counterparts and local populations would be tasked with implementing medium and long term goals. These would build upon development, aid, and security frameworks established by networking with first responders. Continuing the process of interactive policy development they could include civilian organizations and military forces that traditionally work alongside each other on long term development, aid, and security projects. Local populations affected by complex blends of disasters and conflicts would continue to interact and help shape both short and long term emergency aid, development, and security responses. These interactions will offer opportunities to increase civil–military adaptability and improve the ability to anticipate and respond to random events.

Establishing ethical and operational protocols

The new civil–military space will have a range of new and old ethical and operational challenges. Establishing protocols for best practices that do the least harm to survivors and responders is becoming increasingly complicated by the new mix of civil–military actors, new technologies, and new human–technology interactions. While big data and AI have the capability to vet and confirm information sources they cannot detect the nuanced coded language, innuendo, and underlying trends that can predict outbreaks of imminent and near or long term violence. This would be especially important in situations where complex crises are a blend of natural disasters, emergencies, and ongoing conflicts. This approach would allow the emergence of more effective strategies that respond to rapid challenges while taking advantage of small opportunities that have the potential to scale up into larger successful projects. A consistent depth of analysis and thought will need to be applied both in advance of anticipated events and while circumstances are unfolding. Reversing the current emphasis on implementing fixed international aid, development or security programs, these policies can become living interactive documents underpinned by core values and principles that respond to both circumstances and opportunities to support spontaneous peaceful actions and de-escalate violence.

As growing numbers of independent local and international humanitarian actors both in and outside of crisis zones employ a range of rapidly evolving technologies they are adding layers of complexity to defining civil–military relationships and roles. The use of drones is especially problematic as guidelines for legal, ethical, and overall best practices struggle to keep pace. These challenges cut across technical, institutional, and regulatory capacities. Recent deployments of drones in "humanitarian UAV space" accompanied by the corresponding reconfiguration of civilian and military actors in emergency settings have reinforced the potential for UAV/drones to have both positive and negative impacts and disruptive effects (Meier 2015e: Humanitarian UAV Missions).

There are also different legal and cultural interpretations connected to the use of these technologies. The operation of UAVs/drones falls under aviation guidelines and require liaising with a local air traffic control tower similar to flying larger aircraft. This includes filing flight plans, logging conditions and incidents, and other flight-related data. Data privacy guidelines for using and protecting aerial images from UAVs have only begun to be developed but routinely reducing the resolution of images that contain sensitive information before making these photos public is a good general practice. Withholding GPS or any other location information is also important where circumstantial details are unknown or unconfirmed. Most humanitarian and aid organizations are more interested in being given analysis of what the pictures are depicting rather than in seeing the "raw" aerial images. Sharing these photos with local authorities and local communities is important both for helping them identify needs and in gaining their support. Every photo that is published should include contact information so the photographer can be notified if the material is sensitive. UAV/drone operators are also cautioned to avoid practising "drone journalism" where they share images with media outlets that can be misused or interpreted resulting in a backlash or unintended consequences for local communities (Meier 2015e: Humanitarian UAV Missions). Most importantly this technology is readily available to everyone so there are few ways to restrict its use by international criminal and terrorist syndicates who can equally utilize drones to identify locations of enemies or vulnerable groups or to map building layouts for future attacks.

Surging alongside technological innovations are the increasing numbers of non-traditional armed forces operating in international civil–military space, which further complicates negotiation of roles for civil–military responders. These include private security companies as well as locally led security actors hired to deliver humanitarian assistance. Armed non-state actors who include subgroups of armed opposition groups, militias, vigilantes, crime groups, and gangs may also provide assistance during a crisis. While perhaps well intentioned these groups are ungoverned and unaccountable to the country within which they operate. There has been a growing number of international efforts to identify these groups and address trends and challenges presented by their presence during complex crises. While some groups may be apolitical they have the same potential to destabilize and damage a country as an organized insurgency. This is especially true when criminal organizations infiltrate national political power structures to reshape and undermine legal systems that in essence create parallel and separate local legal and security institutions (DCAF and Geneva Call 2015: 21, see Chapter 4).

Back to basics

As fast moving crises and emergencies demand urgent responses they are once again creating the illusion they are somehow separate from lessons of the past. Civilian humanitarian and development organizations project the idea that the basic rules have changed as they embrace disaster focused technology and military forces develop weapons and armor that appear to be invincible. Despite this the inescapable conclusion remains that complex crises require complex thinking that will bridge strategic planning and operational levels. The emergence of AI programs to analyze big data flows and the development of robots to undertake tasks in circumstances too dangerous or inaccessible for humans have done little to change the older core issues in the "new" civil–military space. There are no easy solutions to the civil–military dilemmas presented by old or new war scenarios. Despite the rapid growth of technology backtracking to learn lessons from civilian and military history still matters.

Many international humanitarian organizations accustomed to making fast decisions under pressure have held narrow views that either misinterpreted past experiences or saw history as irrelevant to fast moving contemporary conditions on the ground. Eleanor Davey, author of a 2014 Policy Brief Humanitarian History in a Complex World, urges the inclusion of greater historic perspective in shaping future humanitarian actions by making the case that ignoring humanitarian history negates the value historic knowledge can bring to shaping valuable skills, resources and insights that inform practices and policies. Historic experiences can embrace competing narratives of conflicts and scenarios that help to understand underlying complex processes and circumstances in current humanitarian activities. Taking the time to learn about the successes and failures of earlier efforts provides long term insights that help develop rapid effective short term responses that make a strong case for applying resources (Davey 2014: 3–4).

Fred C. Cuny listed dilemmas still relevant to contemporary issues facing foreign militaries in humanitarian operations in his 1989 speech to UN peace-keeping commanders in Finland. Reviewing a history of the military's role in humanitarian and development initiatives he maintained that military forces must examine and take into consideration the conflicting values and perceptions for needs and requirements expressed by the afflicted government, the local military, relief agencies and the population. Intervening military forces may unintentionally represent the repressive power of the host government in a type of "guilt by association" to local populations. Or they may utilize local civil defense agencies dominated by internal security forces assigned to control not aid their civilian population. Cuny emphasized that one of the most important factors contributing to the success of militaries assisting in humanitarian operations was the ability to assume and maintain a "mantle of neutrality" so that at any time one or more parties in a conflict did not perceive that the intervening military had additional motives for itself or other organizations other than the goal of achieving humanitarian objectives. This required that militaries understand local sensitivities associated with different types of aid and development projects. In a pre-9/11 era this meant working out informal agreements by consulting with the host government for a range of "permissible" activities and with insurgents for activities they would allow. Cuny maintained that peacekeeping forces almost always lost public support and credibility when their humanitarian assistance was perceived by insurgents as a "pacification" effort (Cuny 1989).

There are signs that the international civil–military community is beginning to pay attention. Heeding the difficult lessons of Afghanistan and Iraq senior US military leaders warned in late 2013 against succumbing to the lure of thinking that technological advances in weapons and surveillance have an advantage over or are a substitute for understanding and interacting with local civilians who remain the key to establishing security. They agreed that not understanding the human context before going to war in Iraq and Afghanistan are mistakes they cannot afford to repeat when facing the enemies who are embedded in local populations and who do not use conventional methods to fight. Admiral William McRaven, then commander of US Special Operations Command, emphasized that understanding and dominating the human and geographical terrain is as important now as it was 5,000 years ago and will continue to be for another 5,000 years. While technological advances have created a new type of mobility, sustainment and resupply for smaller numbers of forces it has also created new challenges and problems (Parsons 2013).

International civilian organizations have experienced similar reality checks. While there has been widespread support for an international peacebuilding strategy that promotes developed countries' models of democracy, economics, and rule of law as foundations for peace this ignores the history of early state formation processes experienced by these same nations that were far more nuanced and often violent. While support on the international, national, and local level is important it is domestic politicians and politics that have consistently emerged as key factors in facilitating a peaceful environment. To achieve success local political leaders who are perceived to be credible and legitimate by local populations and able to use patrimonial networks to bring elites and other potential spoilers to participate or at the very least not disrupt peace processes must be brought on board from the onset (Valters *et al.* 2015a: 5).

International military peacekeeping forces can provide safe spaces for reconciliation activities and local security providers can prevent minor disputes by supporting local conflict resolution mechanisms in the short term, but achieving a more sustainable peace requires refocusing international strategies. These include working with but not for elite interests, envisioning support for security institutions as long term evolving processes, and slowly broadening participation in development processes from the beginning of initiatives to include local citizens while simultaneously acknowledging the supporting role of the elites. These should create a framework that supports a long term goal of opening opportunities for increased participation by local citizens (Valters *et al.* 2015a: 5).

Future civil–military responders will need to develop civil–military strategies that balance short term conflict mitigation with long term prevention strategies. These include developing risk management strategies for social media, engaging with non-state actors to gain access to civilians in areas they control, and developing relationships with influential religious leaders and faith-based NGOs to support short and long term prevention of conflicts. Two key lessons from Vietnam are still relevant to future civil–military responses and interventions. One is that core US civilian government and military goals need to be clearly articulated in order to develop international and US civil–military programs that can effectively respond to changing environments. A second is that civilian and military roles need to be carefully defined and redefined as circumstances evolve in dangerous environments (see Chapter 1).

The International Institute for Peace held a conference Lessons from the Past, Visions for the Future: The Middle East After 1914 in Manama, Bahrain, in September 2014. This event sponsored by the government of Norway in cooperation with the Salzburg Global Seminar brought together senior government officials, diplomats, and academics from around the world to discuss and remember the 100th anniversary of World War I and its legacies and impacts on the Middle East. They discussed the effects of the 1916 Sykes–Picot Agreement and the San Remo Conference of 1920, which profoundly reshaped the region's borders bringing with it new political influences and challenges that continue to impact the contemporary Middle East (International Institute for Peace 2014).

The relationship between developing policies and utilizing feedback informed by operational realities during their implementation will be especially important to remember as new challenges to global stability loom large on the horizon. These reinforce the case made throughout this book for strengthening the connection between strategy and tactics to develop the type of resilience and adaptive thinking that will be essential to effectively respond to random combinations of risks while simultaneously taking advantage of unexpected opportunities. There is a great deal of evidence that perseverance in seeking peaceful solutions results in better outcomes. Despite what appears to be uniformly gloomy world news a number of experienced diplomats, negotiators, military forces, humanitarians and other practitioners believe that there is "no such thing" as an irreconcilable conflict "however bloody, difficult or ancient." Nor do they believe that a solution is inevitable or that history will somehow be on the side of peace especially if negotiations are handled badly. Instead a well timed combination of "fighting and talking" that utilizes military force while offering a political escape is considered to be the most effective strategy for dealing with groups who use violence to achieve political goals (Powell 2014). While this may continue to prove true it is time to re-evaluate and broaden this approach as communication and technologically enabled public voices challenge traditional top-down civil–military roles and operations to develop more nuanced sophisticated responses to bottom-up articulation of needs. If there is one international civil–military lesson to learn it is that in every age and in every crisis no matter how complex or fast moving context matters, history matters, and perceptions rule.

Bibliography

Adair, Marshall P. 2013. *Lessons from a Diplomatic Life, Watching Flowers from Horseback*. Rowman & Littlefield Publishers, Inc.

Ajami, Fouad. 2007. *The Foreigner's Gift: The Americans, the Arabs, and the Iraqis in Iraq*. Free Press

Aleinikoff, T. Alexander. 2011. "Foreword." *Forced Migration Review*. FMR No. 38. October 2011. Refugee Studies Centre, Oxford University

Almedom, Astier M. and James K. Tumwine. 2008. "Resilience to Disasters: A Paradigm Shift from Vulnerability to Strength." *African Health Sciences* 8, Suppl. 2008. Editorial: S1–S4. http://www.bioline.org.br/request?hs08057 Accessed 15 August 2014

Alvesson, Mats and Kaj Skoldber. 2009. *Reflexive Methodology, New Vistas for Qualitative Research*. Second Edition, Sage

Andrade, Dale and Lieutenant Colonel James H. Willbanks. 2006. "CORDS/Phoenix Counterinsurgency Lessons from Vietnam for the Future." *Military Review* March–April 2006, No. 2: 9–23

Bacur-Marcu, Hari, Philipp Fluri and Todor Tagarev, Eds. 2009. "Defence Management: An Introduction." *Security and Defence Management Series No. 1*. Geneva Centre for the Democratic Control of Armed Forces (DCAF). www.dcaf.ch

Baker, James A. III and Lee H. Hamilton, Eds. 2006. The Iraq Study Group Report. December 6, 2006. http://www.cfr.org/iraq/iraq-study-group-report/p12184 Accessed 7 August 2014

Beebe, Shannon D. and Mary Kaldor. 2010. *The Ultimate Weapon is No Weapon, Human Security and the New Rules of War and Peace*. Public Affairs Books

Booth, David and Sue Unsworth. 2014. *Politically smart, locally led development*. ODI Discussion paper, September 2014. www.odi.org

Boyle, Joe. 2015. "Islamic State and the idea of statehood." BBC News. 5 January 2015. http://www.bbc.com/news/world-middle-east-30150681 Accessed 5 February 2015

Brigham, Robert K. 2004. "Battlefield Vietnam: A Brief History." Vietnam: A Television History. PBS American Experience, series website companion. http://www.pbs.org/battlefieldvietnam/history/ Accessed 22 June 2014

Bureau of Resource Management. 2011. Fact Sheet FY 2012 State and USAID-Core Budget. 14 February 2011. http://www.state.gov/s/d/rm/rls/fs/2011/156553.htm Accessed 4 October 2011

Capelo, Luis, Natalie Chang and Andrej Verity. 2012. Guidance for Collaborating with Volunteer & Technical Communities. Creative Commons Attribution 3.0. August 2012. www.digitalhumanitarians.com, http://crisismappers.net/ Accessed 20 August 2014

Capps, Walter, Ed. 1991. *The Vietnam Reader*. Routledge

Carayannis, Tatiana, Vesna Bojicic-Dzelilovic, Nathaniel Olin, Anouk Rigterink and Mareike Schomerus. 2014. "Practice Without Evidence: interrogating conflict resolution approaches and assumptions." JSRP Paper 11. Justice and Security Research Programme, LSE, February 2014

Casey-Maslen, Stuart. 2013. "Rebels with a cause? The role of armed non-state actors in the protection of civilians." *Humanitarian Exchange Magazine* Issue 58, July 2013. http://www.odihpn.org/humanitarian-exchange-magazine/issue-58/rebels-with-a-cause-the-role-of-armed-non-state-actors-in-the-protection-of-civilians   Accessed 26 October 2013

CCIC-CCCI. 2007. Canada's whole of government approach in Afghanistan: Implications on Development and Peace-building. Briefing Paper, submission by Canada's Coalition to end global poverty (CCIC) to the Independent Panel on Afghanistan. November 2007. www.ccic.ca

CDA Listening Project. 2009. Field Visit Report, Afghanistan, April–May 2009. CDA Collaborative Learning Projects, Revised August 2010. www.cdainc.com Accessed 13 August 2014

Chan, Margaret. 2015. *From crisis to sustainable development: lessons from the Ebola outbreak*. Women in Science Lecture Series. London School of Hygiene and Tropical Medicine. 10 March 2015. http://www.who.int/dg/speeches/2015/ebola-lessons-lecture/en/

Clark, Helen. 2014. *The Future We Want-Can We Make It a Reality?* Lecture at the Dag Hammarskjold Foundation, Uppsala, Sweden, 4 November 2014. http://www.undp.org/content/undp/en/home/presscenter/speeches/2014/11/04/helen-clark-lecture-on-the-future-we-want-can-we-make-it-a-reality-at-the-dag-hammarskjold-foundation/ Accessed 23 April 2015

Clark, Helen. 2015. *Post-2015 Development Agenda*. UNDP. January 2015. http://www.undp.org/content/undp/en/home/mdgoverview/mdg_goals/post-2015-development-agenda/. Accessed 20 April 2015

Clifford, Clark. 1991. "Why We Did What We Did." *The Vietnam Reader*: 145–151. Routledge

Coffey, Major Ross. 2006. "Revisiting CORDS: The Need for Unity of Effort to Secure Victory in Iraq." *Military Review*. March–April 2006: 24–34

Cuny, Fred C. 1989. Use of the military in humanitarian relief, 1987–89 in Sri Lanka, speech delivered at Niinsalo, Finland (Finnish Defence Forces International Centre-FINCENT military academy for future UN peacekeeping commanders). November 1989. http://www.pbs.org/wgbh/pages/frontline/shows/cuny/laptop/humanrelief.html Accessed 5 August 2013

Cure Violence. 2014. The Model, Understand Violence. http://cureviolence.org/partners/international-partners/ Accessed 16 September 2014

Davey, Eleanor. 2014. "Humanitarian History in a Complex World." Humanitarian Policy Group, Policy Brief 59. Overseas Development Institute. May 2014. http://www.odi.org/sites/odi.org.uk/files/odi-assets/publications-opinion-files/8975.pdf

Davidson, Janine. 2010. *Lifting the Fog of Peace, How Americans learned to fight modern war*. University of Michigan Press

DCAF. 2008. The United Nations and Security Sector Reform: a year on from the Security Council open debate. Presentations at the seminar organized by the United Nations Office at Geneva (UNOG). Slovakia, 4 March 2008. Geneva Centre for the Democratic Control of Armed Forces (DCAF).

http://www.dcaf.ch/Publications/The-United-Nations-and-Security-Sector-Reform-a-year-on-from-the-Security-Council-open-debate Accessed 26 March 2013

DCAF and Geneva Call. 2015. *Armed Non-State Actors: Current Trends & Future Challenges*, DCAF Horizon 2015 Working Paper No. 5. Geneva Centre for the Democratic Control of Armed Forces (DCAF). www.dcaf.ch/Publications

DCAF and ICRC. 2014. Addressing Security and Human Rights Challenges in Complex Environments Toolkit. Geneva Centre for the Democratic Control of Armed Forces (DCAF) and the International Committee of the Red Cross (ICRC)

de Greiff, Pablo and Roger Duthie, Eds. 2009. *Transitional Justice and Development, Making Connections*. Social Science Research Council

Denney, Lisa. 2013. "Consulting the Evidence: How conflict and violence can best be included in the post-2015 development agenda." Shaping Policy for Development. July 2013. Overseas Development Institute. www.odi.org

Denney, Lisa and Pilar, Domingo. 2014. "Security and Justice Reform: Overhauling and tinkering with current programming approaches." ODI Security and Justice Workshop Report. http://www.odi.org.uk/sites/odi.org.uk/files/odi-assets/publications-opinion-files/8895.pdf

Department of Defense. 2011. Office of the Undersecretary of Defense (Comptroller). National Defense Budget Estimates for 2012. March 2011. http://comptroller.defense.gov/defbudget/fy2012/FY12_Green_Book.pdf Accessed 5 October 2011.

Department of Defense. 2012. Sustaining U.S. Global Leadership: Priorities for 21st Century Defense. January 2012. Department of Defense

Department of State—USAID. 2010. Joint Summary of Performance and Financial Information—Fiscal Year 2010. http://www.state.gov/documents/organization/157293. pdf Accessed 9 October 2011

De Waal, Alex and Bridget Conley-Zilkic. 2006. Reflections on How Genocidal Killings are Brought to an End. 22 December 2006. http://howgenocidesend.ssrc.org/de_Waal/ Accessed 8 January 2014

Dixon, Paul. 2009. "Hearts and Minds?" British Counter-Insurgency from Malaya to Iraq. *Journal of Strategic Studies* 32:3, 353–81

Donati, Jessica. 2014. "Exclusive: U.S. to leave more troops in Afghanistan than first planned–sources." Reuters. 25 November 2014. http://www.reuters.com/article/2014/11/25/us-afghanistan-usa-idUSKCN0J91BG20141125 Accessed 16 January 2015

Editors. 2012. "From the Editors." *PD Magazine* Issue 8: Winter 2012: 3–4. publicdiplomacymagazine.org

FMR: Forced Migration Review. 2011. The Technology Issue No 38: October 2011. Refugee Studies Centre, Oxford University

Frerks, Georg, Bart Klem, Stefan van Laar and Marleen van Klingeren. 2006. Principles and Pragmatism, Civil–Military Action in Afghanistan and Liberia. May 2006. Commissioned by Cordaid

Friesendorf, Cornelius. 2010. The Military and Law Enforcement in Peace Operations. DCAF 2010. http://www.dcaf.ch/content/search/?SearchText=schnabel&search=&SubTreeArray=32295&classid=dcaf_publication

Fromkin, David and James Chace. 1991. "What Are the Lessons of Vietnam?" *The Vietnam Reader*. Routledge: pp. 91–9

Galtung, Johan. 1990. "Cultural Violence." *Journal of Peace Research* 27:3. August 1990, 291–305. http://links.jstor.org/sici?sici=0022-3433%28199008%2927%3A3%3C291%3ACV%3E2.0.CO%3B2-6

Gamper, Catherine Desiree. 2014. "Interconnected, Inter-dependent Risks." Background paper prepared for the 2015 Global Assessment Report on Disaster Risk Reduction. September 2014. OECD, Public Governance and Territorial Development Directorate

GAO. 2007. Securing, Stabilizing, and Reconstructing Afghanistan, Key Issues for Congressional Oversight. United States Government Accountability Office Report to Congressional Committees. May 2007. GAO-07-801SP

GFDRR. 2014. Building Social Resilience of the Poor: Protecting and Empowering Those Most at Risk. Background Paper prepared for the 2015 Global Assessment Report on Disaster Risk Reduction. September 2014. GFDRR: Global Facility for Disaster Reduction and Recovery

GHA Crisis Briefing. 2015. West African Ebola crisis. Report Synopsis (Crisis Briefing: Humanitarian funding analysis: 13 January 2015). Global Humanitarian Assistance. 14 January 2015. http://www.globalhumanitarianassistance.org/crisisbriefing/west-african-ebola-crisis

Giffen, Alison. 2013. Community Perceptions as a Priority in Protection and Peacekeeping. Civilians in Conflict, Issue Brief No. 2: October 2013. Stimson Center Civilians in Conflict Project. www.stimson.org/research-pages/civilians-in-conflict.

Ginsberg, Benjamin. 2013. "Why Violence Works." *The Chronicle Review* The Chronicle of Higher Education, Section B: B6-B9. 16 August 2013. chroniclereview.com

Giustozzi, Antonio and Silab Mangal. 2014. "Violence, The Taliban, and Afghanistan's 2014 Elections." Peaceworks No. 103. US Institute of Peace. www.usip.org

Glasius, Marlies and Mary Kaldor, Eds. 2006. *A Human Security Doctrine for Europe, Project, principles, practicalities.* Routledge

Global Humanitarian Assistance Team. 2013. Forgotten crises. Global Humanitarian Assistance. http://www.globalhumanitarianassistance.org/infographics/forgotten-crises Accessed 20 August 2014

Global Humanitarian Assistance Team. 2014. "Annex 1: Forgotten crises." Emerging findings from the forthcoming 2014 Global Humanitarian Assistance (GHA) Report Highlights. http://www.globalhumanitarianassistance.org/wp-content/uploads/2014/06/GHA-2014-highlights-summary-1.pdf Accessed 20 August 2014

GNDR. 2015. *We need a Reality Check! How can we ensure impact at the frontline?.* Reality Check: Impact at the frontline. An Implementation Plan for civil society, to ensure the Post-2015 DRR Framework has an impact at the local level. March 2015. Global Network of Civil Society

GNDR Frontline. 2015. *Everyday Disasters and Everyday Heroes: How Frontline finds out from local people what threats they face.* Organisations for Disaster Reduction. March 2015. Global Network of Civil Society

Gordon, Stuart, Amanda Baker, Alexia Duten and Paul Garner. 2010. Study exploring the evidence relating to Health and Conflict interventions and outcomes. Commissioned by the UK Cross Government Group on Health and Conflict

Gorur, Aditi. 2013. Community Self-Protection Strategies, How Peacekeepers Can Help or Harm, Civilians in Conflict. Stimson Center Civilians in Conflict Project. Issue Brief No. 1. August 2013. www.stimson.org/research-pages/civilians-in-conflict

GPI: Global Peace Index. 2012. Global Peace Index 2012. Sydney: Institute for Economics and Peace. http://www.visionofhumanity.org/wp-content/uploads/2012/06/2012-Global-Peace-Index-Report.pdf

Graham, Jim. 2013. Retired USAID Official, Interview April 2013. Arlington, Virginia

Guterres, Antonio. 2013. Building effective and sustainable partnerships. Keynote Address to the Dubai International Humanitarian Aid and Development Conference (DIHAD). Dubai, United Nations High Commissioner for Refugees, 25 March 2013. http://www.unhcr.org/515d8a789.html Accessed 5 March 2014

Harborne, Bernard and Caroline Sage. 2010. Security and Justice Overview. Security and Justice Thematic Paper, Background paper: World Bank Development Report 2011

Harmer, Adele, Abby Stoddard and Kate Toth. 2013. Aid Worker Security Report 2013: The New Normal: Coping with the kidnapping threat. Humanitarian Outcomes. October 2013. www.humaitarianoutcomes.org, www.aidworkersecurity.org

Hartwell, Marcia. 2005. "Perceptions of Justice, Identity and Political Processes of Forgiveness and Revenge in the Early Post-Conflict Transitions." Case Studies: Northern Ireland, Serbia, South Africa. D.Phil. (PhD) dissertation, University of Oxford, UK. Manuscript hard copy, digital: Bodelian Library, University of Oxford, UK

Hartwell, Marcia. 2006. "Violence in Peace: Understanding Increased Violence in Early Post-Conflict Transitions and Its Implications for Development." Research Paper No. 2006/18: UN-WIDER, February 2006. ISSN 1810–2611 ISBN 92-9190-786-3 (internet version)

Hartwell, Marcia Byrom. *forthcoming* "U.S. Civilian–Military Operations in Unsecure Environments: Learning to Negotiate Shared Space." Chapter in Unity of Mission, The Role of Civilian–Military Teams in Managing Conflict and Supporting Peace: Learning from the Past and Planning for the Future. Jon Gundersen and Melanne Civic, Eds. Air University Press

Harvard Humanitarian Initiative. 2011. Disaster Relief 2.0: The Future of Information Sharing in Humanitarian Emergencies. UN Foundation & Vodafone Foundation Technology Partnership. http://hhi.harvard.edu/sites/default/files/publications/publications%20-%20crisis%20mapping%20-%20disaster%202.0.pdf

Haysom, Simone. 2013. "Civil-military coordination: the state of the debate." Special Feature: Civil-military coordination, Humanitarian Exchange. The Humanitarian Practice Network (HPN) at the Overseas Development Institute (ODI), No. 56. January 2013: 3–4

History Place. 1999. The History Place presents the Vietnam War: America Commits 1961–1964; The Jungle War 1965–1968; The Bitter End 1969–1975. Posted 1999. http://www.historyplace.com/unitedstates/vietnam/index-1961.html Accessed 14 June 2014

Hobsbawn, Eric. 1998. *Uncommon People: Resistance, Rebellion and Jazz*. Weidenfeld & Nicolson

Holt, Victoria, Glyn Taylor and Max Kelly. 2009. "Protecting Civilians in the Context of UN Peacekeeping Operations, Successes, Setbacks and Remaining Challenges." Independent study commissioned by the United Nations' Department of Peacekeeping Operations and the Office for the Coordination of Humanitarian. Available on Reliefweb http://www.reliefweb.int

Hopgood, Stephen. 2014. "The end of human rights." The Washington Post, 3 January 2014. http://www.washingtonpost.com/opinions/the-end-of-human-rights/2014/01/03/7f8fa83c-6742-11e3-ae56-22de072140a2_story.html?hpid=z2 Accessed 5 January 2014

HPG Briefing Note. 2003. Humanitarian Principles and the Conflict in Iraq. April 2003. http://www.odi.org/sites/odi.org.uk/files/odi-assets/publications-opinion-files/4862. pdf

Humanitarian Practice Network. 2010. Operational security management in violent environments. Good Practice Review, No. 8 (New Edition). December 2010. Humanitarian Practice Network at Overseas Development Institute (ODI). www. odihpn.org

ICG. 2015a. Crisis Watch No. 138, "January 2015 – Trends." International Crisis Group. 1 February 2015. www.crisisgroup.org

ICG. 2015b. "Syria Calling: Radicalisation in Central Asia." International Crisis Group Policy Briefing. Europe and Central Asia Briefing No. 72. 20 January 2015. www.crisisgroup.org

ICG. 2015c. "Yemen Conflict Alert: Time for Compromise." International Crisis Group. 27 January 2015. http://www.crisisgroup.org/en/publication-type/alerts/2015/yemen-conflict-alert-time-for-compromise.aspx Accessed 2 February 2015

ICRC and FDFA. 2009. The Montreux Document on Private Military and Security Companies. The International Committee of the Red Cross (ICRC) and Federal Department of Foreign Affairs (FDFA) Directorate of International Law. Section for Human Rights and Humanitarian Law, Switzerland, ICRC. August 2009. http://www.icrc.org/eng/resources/documents/publication/p0996.htm Accessed 7 September 2014

IFRC. 2013. "Strengthening humanitarian information: the role of technology." Chapter 3. World Disasters Report 2013. Focus on technology and the future of humanitarian action. International Federation of Red Cross and Red Crescent Societies. www.ifrc.org

IGRC. 2014. "Emerging risk governance." International Risk Governance Council (IGRC). Summary of Roundtable Discussion. 6 June 2014. August 2014. www. IGRC.org

IISD. 2013. Post-2015 Consultation on Conflict, Violence and Disaster Culminates in High-level Meeting. IISD: International Institute for Sustainable Development. News 13 March 2013. http://sd.iisd.org/news/post-2015-consultation-on-conflict-violence-and-disaster-culminates-in-high-level-meeting/ Accessed April 24, 2015

International Federation of the Red Cross and Red Crescent Societies. 2012. Definition: http://www.ifrc.org/en/what-we-do/disaster-management/about-disasters/definition-of-hazard/complex-emergencies/ Accessed 2 February 2012

International Institute for Peace. 2014. "Lessons from the Past, Visions for the Future: The Middle East After 1914." Manama, Bahrain. 10–11 September 2014. http://www.ipinst.org/events/conferences/details/569-lessons-from-the-past-visions-for-the-future-the-middle-east-after-1914.html. Accessed 19 September 2014

Interviews. 2013. International development practitioners in Washington DC (2011, 2013): The Hague, Netherlands; Oxford, UK (February 2013). Unattributed interviews, discussions, phone conversations conducted in confidentiality with names of interviewees withheld by author.

Iraqi Voices. 2004. "Iraq Voices, Attitudes Toward Transitional Justice and Social Reconstruction." International Center for Transitional Justice and Human Rights, University of California, Berkeley. Occasonal Paper Series, May 2004. www.icti.org/images/content/1/o/108.pdf

Jackson, Ashley. 2014. "Negotiating perceptions: Al-Shabab and Taliban views of aid agencies." Policy Brief 61: August 2014. Humanitarian Policy Group. Overseas Development Institute. http://www.odi.org.uk/hpg

Jackson, Ashley and Simone Haysom. 2013. The Search for common ground: civil–military relations in Afghanistan, 2002–13. HPG Working Paper. April 2013. www.odi.org.uk/hpg

Johnson, David. 2014. "Failure to Learn: Reflections on a Career in the Post-Vietnam Army." War on the Rocks Blog. 24 January 2014. http://warontherocks.com/2014/01/failure-to-learn-reflections-on-a-career-in-the-post-vietnam-army/. Accessed 26 June 2014

Jones, Frank L. 2005. "Blowtorch: Robert Komer and the Making of Vietnam Pacification Policy." *Parameters*. Autumn 2005: 103–118

Kaldor, Mary. 2012. The New Peace: A Lecture by Mary Kaldor. Delivered 28 November 2012. http://sites.tufts.edu/reinventingpeace/2012/11/28/the-new-peace-a-lecture-by-mary-kaldor/. Accessed 8 January 2014

Keen, David. 2012. *Useful Enemies, When waging wars is more important than winning them*. Yale University Press

Kelly, Annie. 2013. "Humanitarian workers unprepared for decades of conflict, warns UNHCR." The Guardian online: theguardian.com. Tuesday 30 April 2013 06.33 EDT. http://www.theguardian.com/global-development/2013/apr/30/humanitarian-workers-unprepared-decades-conflict

Ki-moon, Ban. 2011. Remarks to Security Council's open debate on the Protection of Civilians in Armed Conflict by UN Security-General Ban Ki-moon. 9 November 2011. http://www.un.org/apps/news/infocus/sgspeeches/search_full.asp?statID=1371 Accessed 28 January 2014

Ki-Moon, Ban. 2013. "Foreword III–IV." "A Million Voices: The World We Want, A Sustainable Future with Dignity for All." UNDG Millennium Development Goals Task Force, 2013. United Nations Development Group. http://issuu.com/undevelopmentgroup/docs/f_undg_millionvoices_web_full

Kirwen, Erika. 2015. Security progress in Timor-Leste and Liberia: similar lessons. Blog post 25 February 2015. http://www.developmentprogress.org/blog/2015/02/25/security-progress-timor-leste-and-liberia-similar-lessons Accessed 23 April 2015

Koerne, Brendan I. 2014. "How America's Soldiers Fight for the Spectrum on the Battlefield." Wired. 18 February 2014. http://www.wired.com/threatlevel/2014/02/spectrum-warfare/?utm_source=Sailthru&utm_medium=email&utm_term=%2ASituation%20Report&utm_campaign=SITREP%20JAN%2020%20 2014 Accessed 20 February 2014

Komer, Robert W. 1970. Organization and Management of the "New Model" Pacification Program—1966–1969. Rand Document. 7 May 1970. Declassified 2005. http://www.rand.org/content/dam/rand/pubs/documents/2006/D20104.pdf Accessed 9 June 2014

Kovats-Bernat, J. Christopher. 2002. "Negotiating Dangerous Fields: Pragmatic Strategies for Fieldwork Amid Violence and Terror." *American Anthropologist* New Series 104: 1, 208–22. March 2002. http://www.jstor.org/stable/683771 Accessed 25 July 2011

Krause, Keith. 2007. Towards a Practical Human Security Agenda. Policy Paper-No 26. Geneva Centre for the Democratic Control of Armed Forces (DCAF). www.dcaf.ch

Kubo, Hitomi. 2010. "Operationalising human security, A brief review of the United Nations." New perspectives on Human Security. Malcolm McIntosh and Alan Hunter, Eds. pp. 32–7

Larrauri, Helen Puig and Patrick Meier. 2015. Peacekeepers in the Sky: The Use of Unmanned Unarmed Aerial Vehicles for Peacekeeping. https://letthemtalkdotorg.files.wordpress.com/2015/04/meierandpuig_final.pdf

Levinson, Paul. 2012. "Everyone is a Diplomat in the Digital Age." *PD Magazine*, Winter 2012: Issue 8: pp. 9–12. publicdiplomacymagazine.org

McDonald, Avril. 2004. "The Challenges to International Humanitarian Law and the Principles of Distinction and Protection from the Increased Participation of Civilians in Hostilities." Spotlight on Issues of Contemporary Concern in Humanitarian Law and International Criminal Law. Asser Institute. Centre for International and European Law. http://www.asser.nl/default.aspx?site_id=9&level1=13337&level2=13379#_Toc158269144 Accessed 25 January 2014

MacFarlane, Neil and Yuen Foong Khong. 2010. "Foreword." *New Perspectives on Human Security*, McIntosh, Malcolm and Alan Hunter, Eds. Greenleaf Publishing

McFate, Sean. 2010. "The Link Between DDR and SSR in Conflict Affected Countries." U.S. Institute of Peace Special Report 238. May 2010. www.usip.org

McIntosh, Malcolm and Alan Hunter, Eds. 2010. *New Perspectives on Human Security*. Greenleaf Publishing

MacLeod, Mike Spc. 2009. "Odierno visits Al Anbar area advise and assist brigade." 28 October 2009. 1st Brigade Combat Team, 82nd Airborne Public Affairs, News ID40780. http://www.dvidshub.net/news/printable/40780

Mancini, Francesco. 2014. "2014 Top 10 Issues to Watch in Peace & Security: The Global Arena." 17 January 2014. International Peace Institute. http://theglobalobservatory.org/2014/01/2014-top-10-issues-to-watch-in-peace-a-security-the-global-arena.html Accessed 15 August 2014

Meier, Patrick. 2014a. Establishing Social Media Hashtag Standards for Disaster Response. iRevolution Blog. Posted 5 November 2014. http://irevolution.net/2014/11/05/social-media-hashtag-standards-disaster-response/ Accessed 4 March 2015

Meier, Patrick. 2014b. The Rise of the Humanitarian Drone: Giving Content to an Emerging Concept. Posted 30 June 2014. http://irevolution.net/2014/06/30/rise-of-humanitarian-uav/

Meier, Patrick. 2015a. *Digital Humanitarians, How Big Data is Changing the Face of Humanitarian Response*. CRC Press

Meier, Patrick. 2015b. Drones for Good: Technology, Social Movements and the State. Posted 4 February 2015. iRevolution. http://irevolution.net/2015/02/04/drones-for-good-technology-social-movements-and-the-state/

Meier, Patrick. 2015c. Indigenous Community in Guyana Builds Drones for Good. Posted 5 February 2015. iRevolution. http://irevolution.net/2015/02/05/indigenous-community-drones-for-good/

Meier, Patrick. 2015d. Artificial Intelligence powered by Crowdsourcing: The Future of Big Data and Humanitarian Action. Posted 16 March 2015. http://irevolution.net

Meier, Patrick. 2015e. *Humanitarian UAV Missions: Towards Best Practices*. Handbook, most recent update: 7 June 2015. Version 1.5. Humanitarian UAV Network (UAviators). http://irevolution.net/

Melissen, Jan. 2005. Introduction. *The New Public Diplomacy, Soft Power in International Relations*. Palgrave Macmillian

Menocal, Alina Rocha. 2014. *Getting real about politics: from thinking politically to working differently*. April 2014. ODI. http://www.odi.org/publications/8325-politics-development

Metcalfe, Victoria, Simone Haysom and Stuart Gordon. 2012. Trends and challenges in humanitarian civil–military coordination. A review of the literature. HPG Working Paper: May 2012. Humanitarian Policy Group at ODI (Overseas Development Institute). www.odi.org.uk/hpg

Miller, Sarah Deardorff. 2014. "Lessons from the Global Public Policy Literature for the Study of Global Refugee Policy." *Journal of Refugee Studies* 27: 4, December 2014

Milner, James. 2014. "Introduction: Understanding Global Refugee Policy." *Journal of Refugee Studies* 27: 4, December 2014

National Archives. 2008. "DCAS Vietnam Conflict Extract File record counts by Incident Or Death Date (Year) as of April 29, 2008." Statistical Information about Fatal Casualties of the Vietnam War, Electronic Records Reference Report. http://www.archives.gov/research/military/vietnam-war/casualty-statistics.html Accessed 17 June 2014

New America. 2015. "World of Drones." International Security. http://securitydata.newamerica.net/world-drones Accessed 17 March 2015

Nordstrom, Carolyn and Antonius C.G. Robben. 1995. *Fieldwork Under Fire: Contemporary Studies of Violence and Survival*. University of California Press

OCHA. 2011. The Libya Crisis Map. http://libyacrisismap.net/page/index/1 Accessed 31 January 2012. http://libyacrisismap.net/ Accessed 7 February 2012. Please note: Public access to this map is no longer available and UN OCHA Geneva has closed its call for volunteers for Libya (https://www.onlinevolunteering.org/en/org/opportunity/opportunity_form.html?id=15395) and is redirecting potential volunteers to The Standby Volunteer Task Force, http://blog.standbytaskforce.com/ for current information on crisis mapping in Libya and elsewhere

OCHA. 2013a.What is United Nations Humanitarian Civil–Military Coordination. Civil–military Coordination Section. OCHA, Geneva. November 2013. cmcs@un.org. https://docs.unocha.org/sites/dms/Documents/OOM_CMCoord_11November2013_eng.pdf

OCHA. 2013b. "Humanitarianism in the network age, including world humanitarian data and trends 2012." OCHA Policy and Studies Series. OCHA. 2013. United Nations

OCHA/012. 2014. Hashtag Standards for Emergencies. Think Brief. OCHA Policy and Studies Series. October 2014/012. UN Office for the Coordination of Humanitarian Affairs. www.unocha.org

O'Hagan, Ellie Mae. 2014. "Does social media really bring us closer to the reality of conflict?" The Guardian: theguardian.com. Monday 10 March 2014. http://www.theguardian.com/commentisfree/2014/mar/10/social-media-bring-us-closer-reality-conflict-exploited

Ostensen, Ase Gilje. 2011. UN Use of Private Military and Security Companies: Practices and Policies. SSR Paper 3. Geneva Centre for the Democratic Control of Armed Forces (DCAF). http://www.dcaf.ch/publications

Oxfam International. 2011. Whose Aid is It Anyway? Politicizing aid in conflicts and crises. 145 Oxfam Briefing Paper. 10 February 2011. www.oxfam.org

Parsons, Dan. 2013. Technology Alone Cannot Win Future Wars, Senior Military Leaders Say. *National Defense Magazine* 24 October 2013. http://www.national defensemagazine.org/blog/Lists/Posts/Post.aspx?ID=1314

Patel, Ronak B. and Thomas F. Burke. 2009. "Urbanization-An Emerging Humanitarian Disaster." *The New England Journal of Medicine* 361: 8 August 2009, 741–743. nejm.org Accessed 21 August 2014

Peters, Katie. 2014. "Humanitarian Trends and Trajectories to 2030: North and South East Asia, Regional Consultation." Overseas Development Institute www.odi.org

Powell, Jonathan. 2014. "How to talk to terrorists." The Guardian. 7 October 2014. http://www.theguardian.com/world/2014/oct/07/-sp-how-to-talk-to-terrorists-isis-al-qaida

Pretz, Kathy. 2013. Getting a Handle on Projects That Serve the Underserved. The Institute. The IEEE news source. 4 February 2013. http://theinstitute.ieee.org/benefits/humanitarian-efforts/getting-a-handle-on-projects-that-serve-the-underserved

PRT Fact Sheet. 2009. Provincial Reconstruction Teams Fact Sheet. 15 September 2009. Embassy of the United States, US Department of State. http://iraq.usembassy.gov/iraq_prt/provincial-reconstruction-teams-fact-sheet.html

Ray-Bennett, Nibedita S., Anthony Masys, Hideyuki Shiroshita and Peter Jackson. 2014. "Hyper-Risks in a Hyper-Connected World: A call for critical reflective resonse to develop organizational resilience." Input Paper, prepared for the Global Assessment Report on Disaster Risk Reduction 2015. 6 January 2014

Saferworld. 2014. Conflict and the post-2015 development agenda: Perspectives from South Africa. Briefing. February 2014

Schnabel, Albrecht and Marc Krupanski. 2012. "Mapping Evolving Internal Roles of the Armed Forces." SSR Paper 7. The Geneva Centre for the Democratic Control of Armed Forces (DCAF). http://www.dcaf.ch/Publications/

Schoeller-Diaz, David Alejandro, Victoria-Alicia Lopez, John Joseph "Ian" Kelly IV and Ronak B. Patel. 2012. "Hope in the face of Displacement and Rapid Urbanization: A study on the factors that contribute to human security and resilience in Distrito de Aguabanca, Cali, Columbia." Working Paper Series. September 2012. Harvard Humanitarian Initiative. www.hhi.harvard.edu

Schroeder, Ursula C. 2010. "Measuring Security Sector Governance-A Guide to Relevant Indicators." Occasional Paper No. 20. Geneva Centre for the Democratic Control of Armed Forces (DCAF). http://www.dcaf.ch/Publications/

Shachtman, Noah. 2012. "Special Forces Get Social in New Psychological Operation Plan." Danger Room: What's Next in National Security. 20 January 2012. http://www.wired.com/dangerroom/2012/01/social-network-psyop/#more-69776 Accessed 23 January 2012

Shaddid, Anthony Abboudi. 2009. "In Iraq, the Day After." The Washington Post. http://anthonyshadid.com/journalism/pulitzer-prize/in-iraq-the-day-after/ Accessed 21 December 2012

Shanley, Lea A., Ryan Burns, Zachary Bastian and Edward S. Robson. 2013. "Tweeting Up a Storm, The Promise and Perils of Crisis Mapping." *Photogrammetric Engineering & Remote Sensing* October 2013: 865–79

Shearer, David. 2008. "Guidelines for UN and other Humanitarian Organizations on Interacting with Military, Non-State Armed Actors and Other Security Actors in Iraq." August 2008. DSRSG/HC/RC (July 2008). http://ochaonline.un.org/cmcs/guidelines Accessed 5 February 2011

Sheridan, Mary Beth and Dan Zak. 2011. "State Department readies Iraq operation, its biggest since the Marshall Plan." The Washington Post. 7 October 2011.

http://www.washingtonpost.com/world/national-security/state-department-readies-i raq-operation-its-biggest-since-marshall-plan/2011/10/05/gIQAzRruTL_story.html

Simon, Steven N. 2007. "After the Surge, The Case for U.S. Military disengagement from Iraq." CSR No. 23, February 2007 (with a new Foreword by the author September 2007). Council on Foreign Relations. www.cfr.org

Slim, Hugo. 2008. *Killing Civilians, Method, Madness and Morality in War*. Columbia University Press

Smith, Scott S. 2014. "Last Chance: The International Community and the 2014 Afghan Elections." Peacebrief 169. 14 March 2014. United States Institute of Peace. www.usip.org

SOCOM. 2012. "Consolidated Broad Agency Announcement (BAA) for Special Reconnaissance, Surveillance, and Exploitation and Military Information Support Operations Solicitation." H(2222-12-BAA-SORDAC-IN FY 2012-FY2014:13–14. United States Special Operations Command (USSOCOM)

Stedman, Stephen John. 1997. "Spoiler Problems in Peace Processes." *International Security* 22: 2 (Fall 1997), 5–53

Stepputat, Finn and Lauren Greenwood. 2013. "Whole-of-Government Approaches to Fragile States and Situations." DIIS Report 2013: 25. Danish Institute for International Studies. www.diis.dk Accessed 10 March 2014

Sterman, David. 2015. "Will We Still Call It War?" Edition 71, 26 February 2015. The Weekly Wonk @ newamerica. http://weeklywonk.newamerica.net/editions/will-we-still-call-it-war/ Accessed 3 March 2015

Stoddard, Abby, Adele Harmer and Victoria DiDomenico. 2009. "Providing aid in insecure environments: 2009 Update, Trends in violence against aid workers and the operational response." Humanitarian Policy Group (HPG), Overseas Development Institute. HPG Policy Brief 34, April 2009: 1–12. www.odi.org/resources/docs/4243. pdf Accessed 16 September 2009

Stout, Mark. 2013. "Keep Fighting: Why the Counterinsurgency Debate Must Go On." War On the Rocks Blog. 3 December 2013. http://warontherocks.com/2013/12/ keep-fighting-why-the-counterinsurgency-debate-must-go-on/ Accessed 7 July 2014

Tarzi, Amin. 2013. "Transition in Afghanistan: Lessons from the Past." *Small Wars Journal* 14 June 2013. http://smallwarsjournal.com/jrnl/art/transition-in-afghanista n-lessons-fromthe-past

Thakur, Ramesh. 2010. "Foreword." *New perspectives on Human Security*. McIntosh, Malcolm and Alan Hunter, Eds. Greenleaf Publishing

TRADOC Pamphlet 525-3-1. 2014. "Win in a Complex World, 2020–2040." The US Army Operating Concept (AOC). US Army Training and Doctrine Command (TRADOC). 31 October 2014

Tristam, Pierre. 2014. Glossary: Coalition Provisional Authority, or CPA. http:// middleeast.about.com/od/glossary/g/me071201.htm

Turner, Ted. 2011. "Foreward." Disaster Relief 2.0: The Future of Information Sharing in Humanitarian Emergencies. Harvard Humanitarian Initiative 2011. UN Foundation & Vodafone Foundation Technology Partnership. http://hhi.harvard. edu/sites/default/files/publications/publications%20-%20crisis%20mapping%20 -%20disaster%202.0.pdf Accessed 5 February 2012

UNAMI. 2007. "Fact Sheet." UN Assistance Mission for Iraq (UNAMI). 7 August 2007. http://www.uniraq.org/ici/ICI%202007%20Mid%20Year%20progress%20 Report.pdf

UNAMI. 2008. "UNAMI Fact Sheet." November 2008. http://www.uniraq.org/aboutus/unct.asp

UNAMI Focus. 2009. "SRSG Ad Melkert's First Address to the United Nations Security Council." UNAMI Focus, Voice of the Mission, News Bulletin – Issue 36, August 2009. http://www.uniraq.org/FileLib/misc/FocusAug09EN.pdf

UN Department of Economic and Social Affairs. 2014. World Urbanization Prospects: The 2014 Revision, Highlights. United Nations, New York. http://esa.un.org/unpd/wup/Highlights/WUP2014-Highlights.pdf Accessed 14 October 2014

UN DG Millennium Development Goals Task Force. 2013. "A Million Voices: The World We Want, A Sustainable Future with Dignity for All." United Nations Development Group. http://issuu.com/undevelopmentgroup/docs/f_undg_millionvoices_web_full

UNDP. 2013. *The Global Conversation Begins: Emerging Views for a New Development Agenda*, United Nations Development Group

UNDP. 2014. Delivering the Post-2015 Development Agenda. UN Development programme. 31 October 2014. http://www.undp.org/content/undp/en/home/library page/mdg/delivering-the-post-2015-development-agenda/ Accessed 27 April 2015

United Nations High Commissioner for Refugees. 2007. "Staff safety." Handbook for Emergencies, 3rd edn. February 2007: 520–31. http://www.the-ecentre.net/resources/e_library/doc/ThirdEdition.pdf Accessed 3 October 2011

UN World Conference on Disaster Risk Reduction 2015. Public Forum on Social Implementation of Disaster Robots and Systems. 12–16 March 2015. Sendai, Japan. http://www.ieee-ras.org/educational-resources-outreach/un-symposium

USAID. 1975a. "Development Planning, United States Economic Assistance to Viet Nam, 1954–1975." Viet Nam Terminal Report, prepared by the Asia Bureau. 31 December 1975. US Agency for International Development. Scanned copy of unpublished typed manuscript sent to author

USAID. 1975b. Volume I "A Summary, Part 1." Viet Nam Terminal Report, Vol. I Prepared by the Asia Bureau, Office of Residual Indochina Affairs, Viet Nam Desk. 31 December 1975 (handwritten note by author of Vol I, Robert V. Craig, Sr). US Agency for International Development. Scanned copy of unpublished typed manuscript sent to author

USAID. 1975c. Volume II "Agriculture, A Summary, Part II." Viet Nam Terminal Report. Vol. II, Prepared by the Asia Bureau, Office of Residual Indochina Affairs, Viet Nam Desk. 31 December 1975. US Agency for International Development. Scanned copy of unpublished typed manuscript sent to author

USAID. 2014. History, Who We Are. USAID. http://www.usaid.gov/who-we-are/usaid-history Accessed 17 June 2014

USIP and PKSOI. 2009. "Guiding Principles for Stabilization and Reconstruction." US Institute of Peace and US Army Peacekeeping and Stability Operations Institute. Endowment of the United States Institute of Peace

Valters, Craig, Erwin Van Veen and Lisa Denny. 2015a. Security Progress in post-conflict contexts, between liberal peacebuilding and elite interests. ODI Development Progress, Dimension paper 02. March 2015. developmentprogress.org

Valters, Craig, Sarah Dewhurst and Juana de Catheu. 2015b. After the buffaloes clash, Moving from political violence to personal security in Timor-Leste. Case Study Report, Security, ODI, January 2015. developmentprogress.org

van de Kuijt, Judith. 2012. "The comprehensive approach and NGOs: coherence and its effects on dealing with dilemmas and differences. The cases of Denmark, UK, and the Netherlands in Afghanistan." January 2012. Masters thesis. University of Nujmegan (sent to author by Cordaid)

van der Lijn, Jaair. 2011. "'3D' The next generation: Lessons learned from Uruzgan for future operations." Netherlands Institute of International Relations 'Clingendael'. The Hague. 21 November 2011. http://www.clingendael.nl Accessed 10 March 2013

Van Evera, Stephen. 2007. "The War on Terror: Forgotten Lessons from World War II." *Middle East Policy* XIV: 2, Summer 2007, 59–68

Vergun, David. 2015. "Solarium 2015: Forcing multiple dilemmas on enemy." 2 March 2015. http://www.army.mil/article/143728/Solarium_2015__Forcing_multiple_dilemmas_on_enemy

Verity, Andrej. 2011. "The [unexpected] Impact of the Libya Crisis Map and the Standby Volunteer Task Force." Standby Task Force Blog. 19 December 2011. http://blog.standbytaskforce.com/ Accessed 3 February 2012

Villaveces, Jeffrey. 2011. "Disaster Response 2.0." Forced Migration Review, The Technology Issue: No. 38. Refugee Studies Centre, Oxford University

Visser, Reidar. 2009. "Biden, US Policy in Iraq and the Concept of Muhasasa." 6 July 2009. www.historiae.org

Waddington, Stephen. 2014. "Public Diplomacy in a networked world." Posted Wednesday, 12 February 2014. http://conversation.cipr.co.uk/component/k2/3892-public-diplomacy-in-a-networked-world/3892-public-diplomacy-in-a-networked-world?utm_source=feedburner&utm_medium=twitter&utm_campaign=Feed%3A+WhatsNewInPd+%28What%27s+New+in+Public+Diplomacy%29#When:23:51:11Z Accessed 13 February 2014

Wall, Imogen. 2011. "Citizen Initiatives in Haiti." Forced Migration Review, The Technology Issue: October 2011: FMR 38. Refugee Studies Centre, Oxford University

Warmsler, Christine and Ebba Brink. 2014. "The Urban Domino Effect: A Conceptualization of Cities' Interconnectedness of Risk." Lund University, Centre for Sustainability Studies (LUCSUS), Input Paper prepared for the Global Assessment Report on Disaster Risk Reduction 2015, January 2014

White, Jeremy Patrick. 2009. "Civil Affairs in Vietnam." International Security Program, Center for Strategic and International Studies (CSIS). 28 January 2009. http://csis.org/publication/civil-affairs-vietnam Accessed 23 June 2014

WHO. 2002. "World report on violence and health: summary." World Health Organization. http://www.who.int/violence_injury_prevention/violence/world_report/en/summary_en.pdf Accessed 7 August 2014

WHO. 2014. "Infection prevention and control of epidemic- and pandemic-prone acute respiratory infections in health care." Guidelines. World Health Organization. 2014. http://apps.who.int/iris/bitstream/10665/112656/1/9789241507134_eng.pdf?ua=1

Willman, Alys and Megumi Makisaka. 2010. "Interpersonal Violence Prevention, A Review of the Evidence and Emerging Lessons." Conflict, Crime and Violence Team. Background Paper. Social Development Department (SDV) World Bank. 24 August 2010. World Development Report 2011. http://siteresources.worldbank.org/EXTWDR2011/Resources/6406082-1283882418764/WDR_Background_Paper_Willman.pdf Accessed 10 August 2014

Wooldridge, Mike. 2015. "Boko Haram's IS pledge raises the stakes." BBC News, 8 March 2015. http://www.bbc.com/news/world-africa-31789443

Workshop for Post-Conflict Afghanistan. 2003. "Summary for Workshop on Design of an Evaluation Framework for Post-Conflict Afghanistan." Kings College, London. 31 July 2003. Unpublished workshop summary notes by author and participants

World Bank. 2013. "World Development Report 2014: Risk and Opportunity—Managing Risk for Development." World Bank. doi: 10.1596/978-0-8213-9903-3. License: Creative Commons Attribution CC BY 3.0

World Development Report. 2011. "Conflict, Security, and Development." The World Bank. 2011 IBRD/WB: www.wdr2011.worldbank.org/fulltext

Wulf, Herbert. 2006. "The Challenges to Re-Establishing A Public Monopoly of Violence." In: *A Human Security Doctrine for Europe, Project, principles, practicalities.* Marlies Glasius and Mary Kaldor, Eds. Routledge

Zetter, Roger. 2015. Protection in Crisis, Forced Migration and Protection in a Global Era. Transatlantic Council on Migration, a project of the Migration Policy Institute, March 2015. www.migrationpolicy.org

Ziemke, Jen. 2014. The International Network of Crisis Mappers. http://www.un-spider.org/book/5099/4c-challenge-communication-coordination-cooperation-capacity-development Accessed 26 August 2014

Index